Praise for *The Quantum Moment*

"Rich and entertaining . . . [*The Quantum Moment*] is an introduction to the brave new world we inhabit."
—Amir Alexander, *New York Times*

"Approachable . . . [and] entertaining. . . . Crease and Goldhaber detail in mostly demotic terms and chipper tones the history of quantum physics, its main concepts, its discoverers, its prominent detractors (Einstein among them), and its subsequent jargonization—how and why phrases like *quantum leap* and *uncertainty principle* have escaped the highly technical shackles of the scientific lexicon to become metaphors within literature, philosophy, and self-help."
—Max Ross, *Los Angeles Review of Books*

"The reader will walk away from the book with a deeper appreciation for the men who developed quantum physics, the challenges they faced, and the breadth of impact that their discoveries had on our modern society."   —Andrew Zimmerman Jones, about.com

"Pleasant and informal. . . . *The Quantum Moment* is a good introduction to concepts in quantum theory and will help us better understand how science is bound up with human culture."
—Thiago Hartz, *Physics Today*

"A fascinating tour of some of the most captivating concepts of quantum theory."   —David Kaiser, MIT, author of
How the Hippies Saved Physics

"An amazing book for scientists and humanists alike! Every page yields surprises—not only about the complex history of quantum physics but about how it impacts our understanding of ourselves in daily life. Required reading for anyone concerned with casting the fate of humankind in a radically new light."

—Edward S. Casey, Distinguished Professor of Philosophy, SUNY at Stony Brook, author of *The World at a Glance*

"A delight! A tour de force that is both illuminating and extraordinarily readable."

—Gino Segrè, University of Pennsylvania, author of *Ordinary Geniuses: How Two Mavericks Shaped Modern Science*

The Quantum Moment

The
Quantum Moment

*How Planck, Bohr, Einstein, and Heisenberg
Taught Us to Love Uncertainty*

Robert P. Crease
Alfred Scharff Goldhaber

W. W. Norton & Company
Independent Publishers Since 1923
New York · London

For information about permission to reproduce selections from this book,
write to Permissions, W. W. Norton & Company, Inc.,
500 Fifth Avenue, New York, NY 10110

For information about special discounts for bulk purchases, please contact
W. W. Norton Special Sales at specialsales@wwnorton.com or 800-233-4830

Manufacturing by RR Donnelley, Harrisonburg
Book design by Kristen Bearse
Production manager: Devon Zahn

Library of Congress Cataloging-in-Publication Data

Crease, Robert P.
The quantum moment : how Planck, Bohr, Einstein,
and Heisenberg taught us to love uncertainty / Robert P. Crease,
Alfred Scharff Goldhaber. — First edition.
pages cm
Includes bibliographical references and index.
ISBN 978-0-393-06792-7 (hardcover)
1. Quantum theory—Popular works. 2. Physics—Popular works.
I. Goldhaber, Alfred S. II. Title.
QC174.123.C74 2014
530.12—dc23
2014011427

ISBN 978-0-393-35192-7 pbk.

W. W. Norton & Company, Inc.
500 Fifth Avenue, New York, N.Y. 10110
www.wwnorton.com

W. W. Norton & Company Ltd.
Castle House, 75/76 Wells Street, London W1T 3QT

1 2 3 4 5 6 7 8 9 0

To our students, whose spirit and insight enlivened our course and this book

Contents

The Quantum Moment

Introduction

Not long ago, the two of us were in a classroom listening to our students give their final presentations, which had to consist of original work. Two students read short plays; one performed a hip-hop song; others created artwork in various media. One student, who was being treated for schizophrenia, explained—bravely, haltingly—what it was like to suffer from the disease, using the language of superposed states in quantum mechanics, bringing the typically boisterous and chatty class to dead silence. The next presentation was by a pair of science majors who donned safety goggles, poured liquid nitrogen on a special kind of magnet which is a Type II superconductor, and demonstrated "quantum levitation," explaining the Abrikosov-Meissner effect on which it is based. In between, students made jokes about references to quantum topics that they'd found, including web pages about things like "quantum beer" and "quantum jazz." The last project was by a mechanical engineering major. He brought a stack of odds and ends that he said he had thrown together—Pepsi bottles, eyeglasses, Scotch tape, a Ping-Pong ball—to the front of the room. He astounded the class by shining a flashlight on the stack from a particular perspective—and it projected the clear image of a cat against the wall.

These projects were the final assignment for The Quantum Moment, a class that we've cotaught for half a dozen years. An elective open to physics and philosophy majors, it attracts a diverse range of students. Humanities students who take it are intrigued about what the word "quantum" means. Science students who

Cat by Juan Mesa.

take it want to know whether and why a scientific term can really be applied to human behavior. Other students are drawn by the catchy title or because the course satisfies certain requirements and fits their jammed schedules.

The course is about the cultural impact of the quantum. The quantum, the name for the fact that energy comes in finite packages and is not infinitely divisible, was introduced in 1900 to explain puzzling results in a remote corner of physics about how light is emitted and absorbed. Two quantum revolutions followed. The first occurred between 1900 and 1925, when scientists discussed and developed the theory without attracting much public attention. Then, in 1925–27, the theory was transformed by a second revolution, called quantum mechanics, whose bizarre implications became the subject of public curiosity and discussion. Even today, over eighty years later, humanity is still unsure what to make of the second quantum revolution, finding it puzzling, visionary, and even shocking. New books about the quantum appear almost monthly. The word *quantum* is applied to everyday life in plays, poetry, movies, paintings, fiction, music, philosophy, numerous company names, and "pop science" approaches to psychology and neuroscience. Heisenberg's uncertainty principle, Schrödinger's cat, parallel worlds, and other concepts and images of quantum mechanics show up on T-shirts and coffee cups, in cartoons, novels, poems, and movies. They are found on failblog.org, a showcase of cultural memes. In the American television drama series *Breaking Bad*, "Heisenberg" is the pseudonym of the protagonist, a high school chemistry teacher who manufactures and sells the illegal drug crystal methamphetamine. In March 2012, a *New York Times* opinion article argued that quantum terminology was the correct way to describe the campaign and personality of presidential candidate Mitt Romney: "For we have entered the age of *quantum politics*; and Mitt Romney is the first quantum politician."[1] President Barack Obama, who defeated Romney in the November 2012 election, in interviews has invoked Heisenberg's principle to explain how he seeks guidance from advisers. In the twenty-first century, we ask our class, why does the idea of the quantum still appear as a metaphor in the *human* realm—wild and mysterious, packed with creative force?

Our readings vary from semester to semester but include books and articles on history, philosophy, and sociology, several plays (including Michael Frayn's *Copenhagen*), and have included works of fiction (such as Neal Stephenson's *Anathem*). We provide background for how diverse concepts and images—the uncertainty principle, complementarity, Schrödinger's cat, parallel universes—sprang from the quantum idea. We ask students to think carefully about how quantum language and imagery is used and abused. In the article entitled "How You Get That Story: Heisenberg's Uncertainty Principle and the Literature of the Vietnam War," what insight does this reference add to understanding journalism? Is the use of parallel worlds in movies such as *Another Earth* and *Rabbit Hole* an incisive semiscientific plot twist or distracting ornament?

The students' final projects involve creating, singly or in groups, a quantum-inspired work of their own. In addition to the ones already mentioned, they have also written songs, created multimedia presentations, decorated clothing, and created performance "art" by spouting or spoofing quantum nonsense. One year, students made a video about a second-year physics major (played by a second-year physics major), trying to learn the meaning of quantum mechanics. He goes into Stony Brook's Physics Help Room—an actual place!—hoping a good tutor will be on hand that day. An English major (played by an English major) rushes up to him:

ENGLISH MAJOR. "Thank God you're here! I have to tell somebody! I just read this article on quantum mechanics—you're never going to believe this—with Newton everything was certain, but now with quantum mechanics nothing is certain anymore! There's this Heisenberg uncertainty principle and, I mean, I could go outside and *there's not even an outside.* . . . I left my dorm today and it was sunny and now (pointing to her umbrella) it's probably raining.

PHYSICS MAJOR, *exasperated.* It's a beautiful day outside, I promise.

ENGLISH MAJOR. Everything is uncertain now. With Newton it was fine and now it's not fine anymore. . . . Am I a particle or a wave? Are you even real?

PHYSICS MAJOR, *irritated.* You're fine, I'm fine, your hamster's fine, your dorm's fine.

ENGLISH MAJOR, *suddenly widening her eyes.* My hamster! My dorm! . . . I don't *know* what my hamster is doing right now . . . It's all a mess—

Physics major writes "WHY ME?" in his notebook, stands up, and leaves the room.

"I was reviewing your school's expanded course offerings. 'The Poetry of Quantum Mechanics in the Age of Professional Wrestling' seems a bit contrived."

One of us is a physicist, the other a philosopher. The physicist (Goldhaber) teaches an introductory course on the foundations of quantum mechanics: the now quite complete and well-understood theory of matter and energy based on the discovery that, at the subatomic scale, energy comes in amounts of finite size. This remarkable theory has yet to make a prediction that has been proven wrong. The philosopher (Crease) teaches courses that explore the implications of this theory for concepts such as space, time, causality, and objectivity. Our course came about because the two of us often discussed our bafflement with the way scholarly and popular language is sprinkled with references to "quantum this" and "quantum that," sometimes in the form of genuine insights, at other times of meaningless buzzwords designed to impress or dazzle.

Initially, we assumed the cultural impact of the quantum was largely inconsequential, and that we would mostly find terms and images used by charlatans to impress the gullible, and artists with a dash of scientific literacy who appropriated the cultural authority of science. We found that quantum language and images are indeed often used in humorous, pretentious, wacky, and exploitative ways. But we also noticed that the quantum appeared to provide important new terms and ways of thinking about human beings and the world that have become second nature. We were startled to discover the range of ways in which quantum language and imagery have permeated our world. *Quantum* (plural: *quanta*) is medieval Latin for "a set amount" or "a unit," and stems from an Old Latin adjectve meaning "how much". These terms continue to appear in conventional ways springing from the word's origin, one example being in the title of the still often cited "Quanta Cura," an 1864 Papal encyclical written by Pius IX long before quantum mechanics. The title is taken from the document's first few Latin words, and means "With how much care." But after the development of quantum mechanics, many writers and poets made inventive use of quantum language and imagery. *Fun Home*, Alison Bechdel's graphic memoir of her relationship with her father and of coming out—one of the best books of 2006—includes a terrific scene where her father takes her to New York City a few weeks after the famous Stonewall riots, heralded as initiating the gay liberation movement. "[W]hile I acknowledge the absurdity of claiming a connection to that mythologized flashpoint," Bechdel writes in an evocative sentence, "might not a lingering vibration, a quantum particle of rebellion, still have hung in the humectant air?"[2]

What splintered historical-linguistic path possibly connects the meanings of Pius IX and Alison Bechdel? How is the permeation of art, literature, and philosophy by the language of physics even possible? Has it acted on the social realm, changing our habits and perspectives? In the other direction, have aspects of general culture influenced the development of quantum mechanics? Is there a resonance between developments in the arts and general culture on the one hand, and quantum physics on the other? Why is quantum mechanics, discovered over a century ago by scientists who now take

From *Fun Home: A Family Tragicomic*
by Alison Bechdel.

it for granted, treated like a new discovery by each generation of
writers and artists? Interpreting quantum mechanics, we discovered,
is a kind of hermeneutic Rorschach test in which we learn much
from investigating why some people connect quantum mechanics
with Eastern mysticism and others attribute its creation to the social
anxiety prevailing at the time.

The challenge of figuring out the meaning of quantum
mechanics—how to connect it with the familiar world and its
issues—was one of the great intellectual issues of the early twentieth century. While it may be impossible to instruct nonscientists
in the physics and math of quantum mechanics without extended
training, it is possible to illustrate its conceptual problems or puzzles, and to pursue the connections that artists and writers have

made with these. In what follows, we use brief recapitulations of the development of quantum theory to explain how its terms and images began to appear in art, literature, and everyday language, one after the other, each in its own way. The book unfolds as one long tale consisting of a series of interrelated short stories. After each chapter and before beginning the next, we provide an interlude that gives technical details or mentions related topics that we did not cover in the main text. Some readers might prefer to skip these discussions and may do so without losing pace.

Chapter One

The Newtonian Moment

"I kind of absolutely love the Maybe Dead Cats," I say.
"They're not bad, yeah. A bit pseudointellectual but, hey, aren't
we all?"
"I think their band name is a reference to, like, this physicist guy,"
I say. In fact, I *know* it. I've just looked the band up on Wikipedia.
"Yeah," she says. "Schrödinger. Except the band name is a total
fail, because Schrödinger is famous for pointing out this paradox
in quantum physics where, like, under certain circumstances, an
unseen cat can be *both* alive *and* dead. Not *maybe* dead."

—Will Grayson, Will Grayson

The Austrian physicist Erwin Schrödinger proposed his now-
famous cat image half-jokingly, in 1935, as a comment on his
colleagues' failure to think through quantum mechanics. He could
hardly have imagined that the cat would become firmly ingrained
in popular culture, figuring in novels and cartoons, and embla-
zoned on coffee cups and T-shirts. One recent example crops up
in *Will Grayson, Will Grayson*, a young-adult novel by John Green
and David Levithan (2010). In it, Will asks Jane—a girl for whom
he has mixed and unexpressed feelings—about Schrödinger's cat.
Jane describes the Austrian physicist's famous thought experiment,
and then adds that Schrödinger "was not endorsing cat-killing or
anything . . . just saying that it seemed a little improbable that a cat
could be simultaneously alive and dead."

Will ponders that a moment. Thinking of his own mixed
emotions—he is really attracted to Jane, but declined her first offer

of a kiss—he doesn't find the idea that something can be real and not real at the same time at all improbable. "[A]ll the things we keep in sealed boxes are both alive and dead until we open the box," he broods to himself. "[T]he unobserved is both there and not."[1]

A different usage of Schrödinger's cat image appears in *Blueprints of the Afterlife*, an apocalyptic science-fiction novel by Ryan Boudinot (2012). In it, a protagonist named Abby Fogg keeps showing up in both dead and alive forms after—unbeknownst to her—being programmed to infiltrate another reality. In a morgue one day, she creepily stares at two naked and dead identical versions of herself. "[Y]our selfhood, Abby, has gone into superposition," the forensics director tells her. "It's as if you are both alive and dead simultaneously, and this simultaneity is a self-replicating system in which there are various 'snapshots' of your dead self. Which makes an autopsy pretty dang hard, let me tell you."[2]

Schrödinger's cat is one of many images and terms from quantum mechanics to appear in popular culture. Sometimes these are joking or ironic, while at other times they are serious and offer new perspectives on how human beings look at each other and the world.

To understand how images and terms from a remote corner of physics about how light is emitted and absorbed came to exert such an enormous cultural impact, we must first understand the scientific framework in which the quantum appeared, provided by the work of Isaac Newton just over three centuries ago, and the cultural impact left by this work.

Isaac Newton

In 2004, the New York Public Library mounted an exhibition that was remarkable for its scope and content—and location. "The Newtonian Moment: Isaac Newton and the Making of Modern

Culture" was held in the Library's Main Branch at the epicenter of midtown Manhattan on Fifth Avenue and Forty-Second Street. Visitors entered the Beaux-Arts building by passing between the two famous statues of lions—the one nicknamed Patience is on your left, and Fortitude is on your right—up the marble staircase and through the giant columns into the lobby. Signs then directed you to the magnificent main exhibit hall, where instead of the drab drawings and documents that often comprise library exhibits was a stunning array of materials demonstrating the vast cultural impact of the work of Sir Isaac Newton (1642–1726). [In the Julian Calendar, which Britain still used throughout Newton's life.]

Items included globes, orreries (models of the solar system), and telescopes; sculptures and busts of Newton and his followers; paintings of episodes and experiments in Newton's life; and portraits by Hogarth, William Blake, and others about Newton and his work. One caption invoked the poet Alexander Pope's famous couplet,

Nature and Nature's Laws lay hid in Night;
God said "Let Newton be!" and All was Light.

Other items reflected a less celebratory, more ambivalent judgment of Newton and his work. William Blake's drawing and subsequent painting of Newton show the scientist in an unflattering, even contorted position absorbed by instruments and the figures he's drawing on a scroll, turning his back on the beautiful sea of Nature in which he is submerged. Blake respected Newton's accomplishments, but condemned both his mechanical worldview and the dominance of machines that sprang from it. The extent of Blake's revulsion is apparent in his motto, "Art is the Tree of Life; Science is the Tree of Death."

The exhibition included a first-edition copy of Newton's masterwork the *Principia*, displayed in the United States for the very first time, with Newton's own handwritten corrections for the

Newton by William Blake, 1795–1805.

next edition. Also on display were foreign editions of the *Principia*, a digest version aimed at a female audience, and "Tom Telescope," a series of children's books about Newtonian science. Nearby were paintings inspired by light, gravitation, orbits, the spectrum, the rule of measure and order, and other Newtonian themes.

The exhibition illustrated the revolution in worldview brought about by Newton's work. "[T]he era of the Enlightenment and Revolution," wrote the California Institute of Technology historian Mordechai Feingold in the exhibition's catalogue, "may be viewed as the Newtonian Moment, understood as denoting the epoch and the manner in which Newtonian thought came to permeate European culture in all its forms."[3] In ordinary language, "moment" means a period of time. Physicists give it a technical meaning, associated with rotation; a twisting force or torque, or the resistance of a body to such torque. Historians such as Feingold use the word *moment* in a twofold sense: to mean both the specific period of time in which a radically new development took shape, and the longer period over which its effects were exerted.

Newton's scientific achievements influenced art, literature, philosophy, politics, and religion as profoundly as any artist, writer, philosopher, politician, or theologian of his time. The reason was that Newton's discoveries introduced clarity and intelligibility into an era of confusion and anxiety.

Europe in the 1600s was full of terror and turbulence. Wars raged somewhere almost every year that century; England had three civil wars between 1642 and 1651. People believed that the world worked in mysterious ways, driven by occult powers and enigmatic events. Scriptural texts said that time had a beginning and would have an end, but they did little to help explain the present. Most people could fathom only a little of the world, which seemed like a supernatural organism with several parts. The clearest distinction was between heaven and earth. Each behaved differently; while heavenly things were eternal and unchanging, earthly objects changed by being born or dying, growing larger or smaller, transforming into something else, or moving about. Why did they change? Because of occult forces, some inner potential being activated, or somebody's plan. Kings and clergymen claimed to know God's will, but for almost everyone else the world was untamable, unpredictable, and awesome. Royalty exerted absolute power over political matters, feudal lords and clergy over most aspects of daily life.

Born in 1642, the year of the first English Civil War, Newton was the person who did the most to transform that world. He was a gift to humanity; his mother gave birth to him on Christmas Day, his father—a farmer—had died a few months earlier. As a youth Newton scorned the sports and entertainments of his peers, and passed his time building mechanical things like windmills, water clocks, and kites, and by reading the books in the local apothecary's library. While at Trinity College in Cambridge University

(1661–1665), he mastered on his own the new philosophy, physics, and mathematics that scholars throughout Europe were exploring.

In 1665, the Great Plague hit London, devastating the city by killing one out of every five inhabitants, and Newton retired to his mother's farm in Lincolnshire. The enforced idle time left him free to study uninterrupted, and the germs of many fundamental discoveries came to him in physics, astronomy, optics, and mathematics; he also explored many subjects that today's scientists scorn, such as alchemy, the philosopher's stone, and the Book of Revelation and other Biblical texts. One of his greatest discoveries was calculus, or the mathematics of change—the precise description of situations where adding or subtracting a bit of some quantity x, no matter how large or small, results in the adding or subtracting a bit of some other quantity. Newton used the calculus to build his three laws of motion, a cornerstone of his picture of the world as unified, open, and comprehensible, governed by entirely predictable forces. He helped usher in the view of the world as a giant machine, which needed no God to keep running, and whose operation could be understood not just by kings and clergy but by anyone.

Newton once made the preposterously modest remark that he stood on the shoulders of giants. These mammoth shoulders, however, were part human, part industrial. His way was paved by technological advances of his time: by a growing mechanical knowledge, an upsurge in appreciation for applications of mathematics, increased development and use of machines in mining and industry, and—thanks to Galileo and others—hints of a mechanical rather than supernatural order in the heavens, such as in the behavior of comets.

In 1931, a Soviet physicist named Boris Hessen used a Marxist approach to reach the extreme conclusion that both Newton's standpoint and his scientific ambitions were historically motivated. Hessen was an example of the "externalist" approach to history of science. Externalists view science as like other topics that historians study, as a social product, the outcome of a concrete historical and cultural context. Externalists contrast with internalists, for whom the important forces at work in science history are what scientists do in the lab. For internalists, nature's basic design is already prefigured, like a paint-by-numbers drawing: what scientists do is color it in, so to speak. While social, political, and economic factors may affect which colors get applied first and how quickly the picture is filled in, these factors do not affect the structure of the picture itself. Hessen would have none of that. Hessen began his now famous article "The Social and Economic Roots of Newton's *Principia*" by citing Pope's couplet as a typical example of the tendency to treat historical advances—even scientific advances—as products of divine providence or personal genius. This is wrong, Hessen argued; Marx and Engels have shown that what drives history is the development of relations and powers of production in the service of the ruling classes. Newton's agenda was set not by God, but by social forces seeking pursuit of improved transportation, mining, and warfare. What is now called the Hessen thesis asserts that Newton was the agent of bourgeois industrialized society, expanding and perfecting its efforts to control nature and exploit the working class.[4]

Hessen himself was caught up in the social forces of his own time and place, with a darker outcome. Five years after his article appeared, he was arrested by the Soviet secret police, accused of terrorism, tried, convicted, and shot five days before Christmas, 1936.

Whatever its origins and objectives, Newton's magnum opus, the *Mathematical Principles of Natural Philosophy* (1687), universally known as the *Principia*, changed the world. The single most influential scientific work ever written, it spells out for the first time the laws of motion and of universal gravitational attraction. The world Newton depicted is second nature to us, but was not to many of his peers. Some found it hard to picture, even nonsensical. The earth and the heavens are obviously different places! How could the sun move planets millions of miles away—what was pushing and pulling? Why wasn't Newton investigating the inner causes and mechanical springs that govern the universe? Newton said that he was not going to invent hypotheses about such things, that he would look only at the mathematics—but that's unscientific! As Feingold observed in the exhibition catalogue, "a lengthy process of assimilation" was required before "conversion to Newtonianism was possible."[5] But once human beings became natives in Newtonland, his work influenced every corner of human life and culture, with effects lasting far beyond Newton's lifetime.[6]

It gave birth to a historical "moment," recasting past conflicts to open new possibilities for the future.[7] These turning points are cultural paradigm shifts that change what human beings know and do, and how they interpret their experiences.

The Newtonian world's simplicity, elegance, and intelligibility made it reassuring and beautiful. The earth and the heavens were not separate places made of different stuff but part of a *uni*verse, whose space, time, and laws are single and uniform, the same across all scales. This universe is also homogeneous; it is not ruled by ghosts and phantoms that pop up and disappear unpredictably; everything has a distinct identity and is located at a specific place at a specific time. Things do not change by attaining, actualizing, or intensifying their being. The world is like a cosmic stage or billiard table, where things are pushed and pulled by forces. All space is alike and continuous, all directions comparable, all events caused. To understand how and why things change is to understand how and why they, or their parts, move. For the first time in history, people viewed the world as having a genuine unity and consistency—even a logic!

The Newtonian world was also certain and predictable, characteristics articulated most famously by the French mathematician and astronomer Pierre-Simon Laplace (1749–1827):

We ought then to regard the present state of the universe as the effect of its anterior state and as the cause of the one which is to follow. Given for one instant an intelligence which could comprehend all the forces by which nature is animated and the respective situation of the beings who compose it—an intelligence sufficiently vast to submit these data to analysis—it would embrace in the same formula the movements of the greatest bodies of the universe and those of the lightest atom; for it, nothing would be uncertain and the future, as the past, would be present to its eyes.[8]

During the Newtonian Moment, such an intellect represented the ideal of knowledge for scientists; no esoteric knowledge or supreme authority was needed to understand this world. Human beings could imaginatively picture things happening anywhere and on any scale in the universe as they do at our scale. You could model the movements of the objects in the solar system with balls and rods; after the birth of atomic theory hundreds of years after Newton's death, you could do the same with atoms and molecules. Mechanical models could all be expressed mathematically. "I never satisfy myself until I can make a mechanical model of a thing," the British scientist Lord Kelvin said. "If I can make a mechanical model I can understand it. As long as I cannot make a mechanical model all the way through I cannot understand."[9] The laws of motion could also be captured mathematically. The force of gravitational attraction between two bodies, for instance, is given by a single expression: a constant multiplied by the product of the masses of the bodies divided by the square of the distance between them. This mathematical expression holds regardless of what, where, and how far apart the bodies are. Understanding such laws allows us to send people to the moon, and spaceships to the edges of the solar system.

Finally, the Newtonian world can be understood without reference to human actions. For Aristotle, explaining motion required knowing the mechanical pushes and pulls, but also the "why" of what was happening: human purposes and goals. Why is that cart moving down the road? Partly because the horse pulls it and the road resists, but also because the merchant needed to sell wares at the market to support his family and so hooked it to the horse, and so on. But for Newton, the cart's motion is to be explained solely in terms of forces and masses, without references to human motion— or, if human motions are involved, these too are ultimately to be explained by mechanical pushes and pulls. The Newtonian world is a stage on which the only things that move are masses, and the only things that move them are forces. We can study it like the contents of a fishbowl: from the outside without disturbing the contents, as if we observers were separate from what we observe. Measuring may disturb what we measure, but that disturbance can be anticipated and in principle reduced to whatever minimum amount we need. Scientists can "take themselves out" of their measurements, remove themselves from the knowledge they acquire. The Newtonian Moment marked an era in which scientists could happily study the laws of nature mechanically and mathematically, without reference to transcendent purposes, plans, and designs.

Historians Betty Dobbs and Margaret Jacob write that the Newtonian Moment provided "the material and mental universe— industrial and scientific—in which most Westerners and some non-Westerners now live, one aptly described as modernity."[10] It undermined superstition—belief in special, animistic forces and powers. It implied a new role for God. Newton, a believer, assumed that his machinelike universe required a supreme lawgiver and that his descriptions of the world's unity, design, and rationality would foster greater belief in God. This backfired. His description of the regular, mechanical operation of the universe drew attention to its laws, away from its creator. God himself seemed bound by his own laws, a machine-builder whose presence was superfluous once his creation was commissioned. The fact that all "why" questions had mechanical and mathematical answers put religious believers on the defensive.

Nevertheless, Newton's influence was strong and decisive on nearly all aspects of culture, from literature and music to political theory, philosophy, and theology.[11]

Newton's achievement influenced political theory because it favored stable political orders and the modern idea of democracy, weakening arguments for absolutism. Political thinkers were encouraged to seek the laws governing the human world the way Newton had sought the laws governing the physical world.[12] One of Newton's assistants, John Desaguliers, wrote a poem, "The Newtonian System of the World, the Best Model of Government," extolling the orderly solar system as a model for government. All the US founding fathers read Newton and relied on him in their writings about human affairs. Thomas Jefferson clearly had Newton's work in mind when he referred to the "Laws of Nature" in the very first line of the Declaration of Independence. Newtonian-inspired political scientists, as the American philosopher Richard Rorty has characterized them, are those who center social reforms around "what human beings are like—not knowledge of what Greeks or Frenchmen or Chinese are like, but of humanity as such."[13]

Newton's work also gave philosophers a new task. Many philosophers noticed that Newton's work depended on certain ideas— such as infinitely extended and divisible space and time, and universal causality—that are neither self-evidently true like mathematical truths on the one hand, nor are things found in experience on the other. Why do these ideas seem so hardwired? The German philosopher Immanuel Kant showed that such ideas are conditions for the possibility of experience itself, without which no human consciousness would be possible. They are parts of our mental software, which organizes data from the senses in a form

The Sir Isaac Newton, 84 Castle Street, Cambridge, England.

that gives our experience its order and coherence, and Newton's laws their validity. Just as Newton said that he would not venture hypotheses about the inner causes of things like gravity, Kant said he would not venture hypotheses about what things were apart from how we experience them, what he called *noumena* as opposed to *phenomena*.

Poets, painters, and other artists were either inspired or provoked by Newton's work. Romantic poets split between those who believed, with the English poet John Keats, that "the rainbow is robbed of its mystery" by explanations of optical refraction, and those who believed, with his countryman, the poet James Thomson, that one could learn to see rainbows and sunsets with "Newtonian eyes" and "declare, how just, how beauteous the refractive law."

Finally, Newton's achievement exerted an almost cultlike fascination on the public; it provided insight into the operations of the universe that previously had been reserved for religious authorities and mystics. His work was not only discussed in sermons, but also by secular groups of scholars and ordinary people.[14]

Newton's influence is evident in every aspect of modern life, both scientifically and culturally. It is evident in speedometers, gravimeters, altimeters, and other instruments that measure continuously varying quantities. On the outskirts of the city of Cambridge, England, next door to the Lyndsey McDermott hair salon on Castle Street, is a pub called the Sir Isaac Newton. An ordinary joint with an ordinary range of beers, it also offers an ordinary range of food. Why the name? Patrons are likely to stare at you blankly, muttering something about British greatness, history, or the fact that Newton was educated at the university down the road. But the pub's name reminds us that Newton remains a popular icon, too. (For a reminder closer to home, check out the Wikipedia page "Isaac Newton in Popular Culture.") Newton's name has been given to Cambridge University Library's online catalogue, to a unit of force and a temperature scale, to a computer operating system, and to an orbiting X-ray observatory.

The enduring influence of Newton's work is related to the fact that it provides the scientific framework for the way our perceptual apparatus has evolved. We are adapted to a world in which the assumption of continuity is rewarded; we fill it in when we do not actually see it. Magicians know this and call it the law of good continuity.[15] It is how they dupe us by setting things up so that we perceive an action or event that hasn't actually occurred. In other words, magicians manipulate the order that, on Newtonian grounds, we assume to exist in our environment. The Newtonian Moment did not establish this principle—it was there already—but provided a scientific grounding for it. Magicians would have a tough time in the quantum world. The British biologist Richard Dawkins wrote, "Our brains have evolved to help us survive within the orders of magnitude of size and speed which our bodies operate at." These are our comfort zones of intuition, which he calls the "middle world . . . the narrow range of reality which we judge to be normal, as opposed to the queerness of the very small, the very large and the very fast."[16]

In codifying the rules governing our comfort zones of intuition in the natural world, Newton's work also promoted expectations of the human world as also ruled by causality, consistency, and continuity—an expectation that was to have a strong influence on art, literature, and philosophy. Such expectations shape the intuitions, for instance, of the writers and readers of novelistic accounts of people in their unfolding interactions, the novel being a literary form that began to emerge during Newton's lifetime. To be sure, the real lives of most people were anything but smooth, continuous, and law-governed—but the Newtonian vision of the universe provided a model that encouraged people to view their lives as though, at bottom, they were.

Newton's work, in short, influenced nearly every aspect of human life, helping to provide human beings with answers to what Kant said were the three fundamental philosophical questions: What can we know? How should we act? What might we hope for? The Newtonian Moment even has a more intangible, spiritual dimension, which we find articulated by the German-born philosopher Rudolf Carnap and his colleagues in their manifesto of logical positivism entitled "The Scientific Conception of the World." The manifesto was written shortly after the quantum revolution (in 1929), when it was not yet clear how thoroughly its cultural impact would undermine the Newtonian spirit embodied in the manifesto, and the document's spirit is decidedly Newtonian. The scientific world conception, the positivist philosophers wrote, is not so much a set of theses, but an attitude and a direction. "The goal ahead is *unified science*," they wrote, from which "springs the search for a neutral system of formulae, for a symbolism freed from the slag of historical languages, and also the search for a total system of concepts." This is more than a scientific aim, they continued, and implies that a fully rational human being should possess a particular attitude and behavior:

> Neatness and clarity are striven for, and dark distances and unfathomable depths rejected. In science there are no "depths"; there is surface everywhere: all experience forms a complex network, which cannot always be surveyed and can often be grasped only in parts. Everything is accessible to man; and man is the measure of all things. . . . The scientific world-conception knows *no unsolvable riddle*.[17]

The spirit of this positivist manifesto is perhaps the high water mark of the Newtonian Moment.

The Quantum Ambush

The Newtonian Moment lasted for about 250 years. Then, in 1900, it was ambushed. While studying an apparently remote corner of optics, physicists found that they could explain what was happening only by introducing a strange, entirely novel idea: the quantum. The idea was deceptively simple, but would turn out to smash the foundations of the Newtonian world.

The attack unfolded in two phases. In the first, from 1900 to 1925, scientists gradually came to see that the quantum did not fit smoothly into the Newtonian world, and clashed with assumptions about space, time, and causality. But many hoped to find a way to tame their unruly child and find it a place within the elegant Newtonian universe, as they had found place for so much else since the *Principia*'s publication. During this period, the public viewed quantum theory as akin to relativity; terribly important but incomprehensible to all but the brightest scientific lights, who were excited about it for reasons far too complicated to understand.

Then everything changed. In an astonishingly short period, 1925 to 1927, hopes to tame the quantum were dashed by a new theory with foundations fundamentally different from those of classical physics, and whose capstone was the uncertainty principle. Quantum mechanics, as this second revolution was called, explained phenomena that Newtonian mechanics did not, and answered a vast array of questions such as how the sun shines and the dynamics of atoms. It replaced Newtonian physics as the theory of the microscopic realm. In quantum history's second phase, one could speak of the quantum "world" in contrast to the Newtonian world, as though the two referred to different territories governed by different laws. This new territory was exotic and even magical in contrast to Newton's. But it lacked simplicity and elegance due to several troubling features. One was difference across scales; scale matters, for different laws apply to microworld and macroworld. Another was inhomogeneity; some things have a different kind of presence in the world than others. A third was discontinuity; the values of properties like space and time do not flow into each other smoothly as with Newton. A fourth was uncertainty; some properties of Newton's universe, such as position and momentum, could not be simultaneously pinned down or even said to be real. A fifth was unpredictability, and a sixth the inability to take oneself out of certain kinds of measurements; certain properties of microworld phenomena seemed to depend upon being measured for their very existence—thus demolishing a classical conception of objectivity, that all properties have fixed values apart from their being measured. Scientists, that is, found they could not remove themselves from parts of the knowledge picture.

The discovery of the uncertainty principle in 1927 exploded lingering hopes of returning to a deterministic, Newtonian world. A foundational structure of quantum mechanics, the principle implied that it is impossible to know simultaneously the position and momentum of a particle. It meant that Blake's Newton could not learn everything he thought he could about the beautiful sea of Nature he's turning his back on. It meant that Laplace's all-knowing, ideal intellect could not even get started. The principle also helped to change the meanings of many words, including *uncertainty, randomness, chance, cause,* and *probability.*[18] A few scientists, Einstein the most famous, dug in their heels, hoping for a restoration of the Newtonian world. Some scientists tried to explain what was happening in language accessible to nonscientists, and the public found these popularizations fascinating. This spread the struggle far beyond physics, where it began, into ever-widening cultural spheres. Philosophers, artists, novelists, and poets were soon adopting quantum terms and concepts, which began to appear even in everyday language. Cultural forms that had been nurtured by the Newtonian Moment were threatened, opening the door for new forms to evolve.

What did the quantum world mean? Was it like a previously undetected region in some unexplored mountain range, well worth learning about because of its sweeping landscape and bizarre creatures, but with few implications for our own world? Was it like making contact with an alien civilization in some distant galaxy, bound to change how we humans think of ourselves and our position in the universe? Or did Quantumland, which after all lies at the foundation of the world, have still deeper implications for what we consider rational and irrational, real and unreal?

Even after all this time, the Quantum Moment is murkier and more unsettled than the Newtonian Moment it succeeded, and its shape is still being worked out. Human beings have not yet become "quantum natives." A century on, we remain immigrants in Quantumland, still seeking to make this quirky and mind-bending world familiar and to understand its implications. We find quantum mechanics alluring because it seems to speak to difficulties we often face in describing our own experience. Quantum mechanics is strange and so are we.

"Nobody understands quantum mechanics," the physicist Richard Feynman famously said. He was being provocative as usual, but his comment underscores the notorious difficulty and utterly counterintuitive nature of the subject. Why so difficult?

One possible answer is: We can't go native. Evolution has adapted us to think and act in a world whose dimensions and timescales are Newtonian, and quantum effects are not directly noticeable. We have adapted to this world as a species, matured in it as individuals, and are conceptually moored in it by how our minds process experience. In this view, an unbridgeable chasm separates the classical and quantum worlds, so we will always find the latter exotic and incomprehensible. The pleasure in learning about the quantum world is like that of a magic show, where we delight in the mismatch between what we expect and what happens. The thrill will never wear off.

But there's an alternative: that the weirdness stems not from the quantum world but from us. Things are weird only in contrast to the familiar. If what we thought was familiar turned out to be a fantasy and to contain false assumptions—if our world is odder than we think—the quantum world to which we contrast it would not seem so freakish. Blake's Newton might not have been so surprised at the new developments if he had stood up, turned around from his scroll, and consulted his own experiences every so often. Our quest to learn about the quantum world might turn out to have some sort of upside-down Wizard of Oz ending, in which we suddenly realize that what we thought was home was really only a dream, and our world was always a little Oz-like. Then we'd be natives.

The Grand Design

Newton "is our Columbus," wrote Voltaire, for "he led us to a new world."[19] It is a special kind of world, a kind of skeletal version of our own. The only things in it are masses, the only thing these masses do is move about, and the only things that start or stop these motions are forces. Yet Newton showed that these most elementary assumptions produce a vision of the universe that is so simple, elegant, and comprehensive that it is called the "Grand Design."[20]

Newton showed us the Grand Design in his *Principia*. Its basic concepts are force, mass, and velocity. Each mass has a specific position at a specific time. If it moves, its rate of change of position with respect to time is called its velocity. If its velocity changes, its rate of change of velocity with respect to time is called its acceleration. When masses accelerate, they do so because of the influences of forces. These forces arise from interactions between bodies, either from contact or attraction/repulsion. The *Principia* set forth three laws of motion that, in the Grand Design, are held to be valid throughout the universe on all scales and in all regions:

FIRST LAW: A mass that is not acted on by a force stays at rest, or continues to move in a uniform, straight-line motion. If a mass starts to move, or changes its uniform, straight-line motion, it is being acted on by a force.

SECOND LAW: The acceleration of a mass is proportional to the force that acts on it, and in the same direction. In the *Principia*, Newton wrote this out in words. Only a hundred years later would it be reduced to symbols and put in the form of the now familiar equation $F = ma$.

THIRD LAW: The third law applies to the interaction between bodies, and says that each force has a counterforce. As Newton put it, "To every action there is always opposed an equal reaction: or the mutual actions of two bodies upon each other are always equal, and directed to contrary parts." When a horse pulls a stone tied to a rope, Newton continued, the horse is pulled back to the stone as much as the stone is pulled toward the horse.

To predict the future motion of a mass, or deduce its past position, involves working out the rule for calculating at any time the mass's velocity—the rate at which its position is changing—from these three laws. Given that velocity, one may calculate how far the object moves in any time interval. That distance is simply the magnitude of the time interval multiplied by the average velocity during the interval. To determine velocity, Newtonian mechanics introduces another quantity, momentum, which is given by mass, also called "inertia," multiplied by velocity. The bigger the mass, the less easily the object changes its velocity under the action of a force. The average force over a time interval, multiplied by the duration of the time interval, gives the change of momentum during that interval. Thus, to predict all motions, one needs only to know the positions and velocities of all objects at a given time, and to know the forces acting among the objects (as functions only of the positions and velocities) at all times.

Newton set out a fourth law of motion—the GRAVITATIONAL LAW—that governed the physical force (attraction) between bodies. This force is given by a single expression: a constant multiplied by the product of the masses of the bodies divided by the square of the distance between them:

$$F = G\frac{m_1 m_2}{r^2}$$

This expression holds regardless of what, where, and how far apart the bodies are.

In what is known as Newtonian or classical mechanics, these assumptions are elaborated in different forms to apply to many different circumstances—to masses that are solid, liquid, and gaseous; to circular, rotational, and vibrating bodies; to motions of projectiles, pendulums, and waves; to complex systems of many masses. They have been applied to atoms and galaxies. Newtonian mechanics provides a way to capture mathematically the motions involved, no matter how simple or complex.

The Newtonian Moment emphasized the continuities of nature, though now and then a few renegade scientists proposed that discontinuities and disorder lay at the heart of nature. Newton—and his archrival Leibniz, who came up with the same idea at about the same time—created the calculus, the principal mathematical tool of the Newtonian world, on the notion of infinitesimally small quantities. He assumed that the nature to which that tool was applied behaved likewise. "Natura non facit saltus," Newton declared, using an ancient saying that means "Nature does not make jumps." If it did, scale would matter, and the very application of the calculus—and with it, the scope of classical mechanics—would be limited.

The Newtonian world is an austere, abstract world, devoid of human interests and idiosyncrasies. It is a world where all chandeliers, trapezes, and swings are pendulums, all sport and dance instances of $F = ma$, all balls elastic, and all planes go on forever. No matter where you move about in this world—and even if you grow bigger or smaller—the laws always remain the same. If you want to find out what has or will happen in this world, here's what you do: quantify the positions, speeds, masses, and forces; apply the appropriate laws; and turn the calculational crank. The crank relies on a special kind of tool called differential equations, which are a straightforward way to state the relations between continuously varying properties, such as (in the Newtonian world) between velocity and position, or force and momentum. In that language, a small change in x or position is symbolized by dx, and a small change in t or time, dt, with the changes approaching zero. The statement that velocity is the rate of change of position with respect to time is therefore $v = dx/dt$. Similarly, the effect of force to produce acceleration is given by $F = dp/dt$, with momentum $p = mv$. An extremely simple and important ideal example is a system of a point mass attached to a light spring, so that all the inertia may be ascribed to the mass: $dx/dt = v$, $dp/dt = -kx$. Here k is the spring constant, which gives the proportionality between displacement of the mass from the equilibrium point and the opposing force of the spring.

This is called an ideal spring (or Hooke's Law spring, after Newton's nemesis Robert Hooke), for any real spring only stretches a certain amount before the restoring force weakens, no longer increasing proportionally to displacement. Also, any real spring involves a certain amount of friction. Still, an ideal spring is often a good approximation, and its solution—its motion is the same as the projection on one axis of motion around the circumference of a circle with constant speed—is ingeniously simple: the displacement changes with time like a trigonometric function, the sine of an angle proportional to the time since it started with zero displacement.

In the centuries that have elapsed since the *Principia*, scientists discovered many new phenomena that had to be fit into the Grand Design. Electricity was one. The French scientist Charles Coulomb and others studied electrical attraction and repulsion by

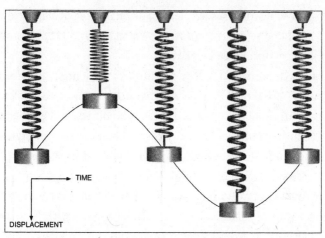

TIME

DISPLACEMENT

Hooke's Law spring, illustrating the sinusoidal motion versus time
of a mass attached to an ideal (massless) spring.

imitating Newton's universal law of gravitation. They found that
the force between two electric charges, or two magnetic poles,
is proportional to the product of the charges, and again inversely
proportional to the square of the distance between them. But while
all masses are positive, electrical charges or poles can be either
positive or negative—meaning that while gravitational masses all
attract each other, like charges or poles repel and unlike attract.
The concept of a field—of a quantity that has a specific value for
each location in space and time—now vastly extended the power
of Newtonian mechanics. Other scientists were able to show how
currents of electric charge produce circulating magnetic fields,
while changing magnetic fields induce circulating electric fields,
establishing a link between electricity and magnetism. The British
scientist James Clerk Maxwell realized that changing electric fields
also induce circulating magnetic fields. When he then put everything known about the new phenomenon called electromagnetism
together, he got an astonishing result: light was nothing more or
less than waves of electromagnetic fields! All these developments
fit easily in Newton's formulation of mechanics, with forces deter-

mining how fast masses accelerate, or change their velocities. All the new phenomena fit comfortably in the Grand Design, and are described by differential equations.

It was not entirely a clockwork universe; scientists learned that the world contained behaviors that looked random. Scientists studying areas such as geology, thermodynamics, and gas behavior developed methods to handle irreversible actions, and statistical tools to handle uncertainty.[21] But this involved what philosophers call epistemological uncertainty—uncertainty in what we know about the objects of study—rather than ontological uncertainty, or uncertainty in nature itself. As Laplace put it with Newtonian confidence, "The word 'chance,' then expresses only our ignorance of the causes of the phenomena that we observe to occur and to succeed one another in no apparent order."[22]

When the quantum first appeared in 1900, it did so in Newtonian territory; it was only discovered because it accounted for a deviation from Newtonian expectations concerning how light behaved in special circumstances. The quantum world could only have been found inside a Newtonian one. For a while, scientists fully expected that it, too, could be made to fit somewhere in the Grand Design, just as had every other phenomenon that they had discovered up to then. They were in for a shock.

Chapter Two

A Pixelated World

In 1967, the critic and novelist John Updike wrote a brief reflection on the photographs and amateur films taken in Dealey Plaza in Dallas, Texas, on November 22, 1963, in the few momentous seconds when President John F. Kennedy's motorcade drove through and he was hit by an assassin's bullets. The more closely and carefully the frames are examined, Updike noted, the less sense they made. Who was the "umbrella man," with an open umbrella despite it being a sunny day? Or the "tan-coated man" who first runs away, then is seen in "a gray Rambler driven by a Negro"? What about the blurry figure in the window *next* to the one from which the shots were fired? Were these innocent bystanders or part of a conspiracy? Updike wrote:

> We wonder whether a genuine mystery is being concealed here or whether any similar scrutiny of a minute section of time and space would yield similar strangenesses—gaps, inconsistencies, warps, and bubbles in the surface of circumstance. Perhaps, as with the elements of matter, investigation passes a threshold of common sense and enters a sub-atomic realm where laws are mocked, where persons have the life-span of beta particles and the transparency of neutrinos, and where a rough kind of averaging out must substitute for absolute truth.[1]

Years later, many frames turned out to have rational explanations. The "umbrella man" was identified—to the satisfaction of all but diehard conspiracy theorists. He testified before a Congressional committee that he had been simply protesting the Kennedy family's dealings with Hitler's Germany in the Second World War, a black umbrella—British Prime Minister Neville Chamberlain's trademark accessory—being a symbol for Nazi appeasers. Far from heralding a breach in the rationality of the world, the umbrella man was just a heckler.

But Updike's description rings true. He knew that when scientists look at the subatomic world frame by frame, so to speak, what they find is discontinuous and strange, its happenings random except when collectively considered. He also knew that ordinary human beings tend to find our lives following a similar crazy logic, even if metaphorically. Our world does not always feel smooth, continuous, and law-governed; close up, it often feels jittery, discontinuous, and irrational. Today's world does not have the gentle geometry of the Newtonian universe, but is more like the surface of a boiling pot of water. Calling that a quantum condition may be scientifically incorrect, but to Updike—and to numerous other writers and poets—it was metaphorically apt.

How did the gentle geometry of the Newtonian world, stable and compelling for over two centuries, come to be perforated by gaps, inconsistencies, warps, and bubbles?

Max Planck

To answer requires a brief detour into the life of Max Planck (1858–1947), who introduced the notion of the quantum into science.

Temperamentally the most conservative of men, Planck had no intention of being a revolutionary. He came from a family of ministers and lawyers: responsible, exacting, honest people. He learned to keep a low profile to survive tumultuous times. When he was six years old in his hometown of Kiel—then in Schleswig-Holstein, now in Germany—he watched as conquering foreign troops marched

Max Planck (1858–1947).

through; when he was fifty, living in Berlin, he watched as Germany lost the First World War; and when he was seventy-four, he was in Berlin when the Nazis seized power. He survived numerous personal tragedies. His oldest son was killed in the First World War, his second son was murdered by the Nazis in the Second, and his house was destroyed in an Allied bombing raid. While not Jewish, during the Nazi terror he was attacked as a "white Jew" for teaching the heretical views of Jews like Einstein. But Planck managed to survive in Germany through the war by remaining low-key.[2]

Scientifically, Planck was also conservative. As a youth he was strongly attracted to science because it allowed him to study a world that was absolute—independent of human actions. Science offered a marvelous refuge from the threatening world around him. "[T]he quest for the laws which apply to this absolute appeared to me," he recalled at the end of his life, "as the most sublime scientific pursuit in life." While he was a student in Munich, one of his professors discouraged him from this sublime pursuit, telling him that, in physics, "almost everything is already discovered, and all that remains is to fill a few holes." No matter; the young Planck was content to fill in those holes and to sweep up dusty corners. As a graduate student in Berlin, his interests focused on thermodynamics, a subfield of physics concerned with the relations between heat, light, and energy. Once again his professors were discouraging, telling him that the field was nearly complete, and again he ignored this advice. Planck had no desire to make discoveries, just to solidify foundations. "A conservative in the root meaning," his biographer wrote, "his particular effectiveness lay in his ability to adapt to, and even direct, current realities while saving, and acting on, traditional values."[3]

How ironic that this man, who intended to tidy up science's loose ends, would introduce an idea that would strike at its deepest foundations! He did this in the course of his work at the Physikalisch-Technische Reichanstalt, the German Bureau of Standards, which had been established in 1887 as the first national laboratory of weights and measures in the world. The needs of the new electric industry had been growing by leaps and bounds, and the German government had asked the Reichanstalt to create a rating system for lightbulbs, for which its scientists needed to analyze what happened when materials absorb all the light falling on them, and then re-emit the light in the most even distribution of colors allowed by the laws of physics. Materials in that condition—the best absorbers of incoming light—were baptized "black bodies" by Planck's teacher Gustav Kirchhoff (1824–1887). The German government therefore asked scientists at the Reichanstalt to investigate black-body radiation, and to produce a formula describing its "normal spectrum," as the profile was called, or how the intensities and frequencies of the radiation vary with temperature.

Planck, who in 1892 succeeded Kirchhoff as a professor in Berlin, liked this problem. For one thing, he would be fulfilling his duty as a civil servant by working on a topic in the national interest. For another, he was prepared, for the problem was closely connected with his previous studies. Finally, the fact that the glow depended only on the temperature of the material and not on its chemistry suggested that the solution would be fundamental—akin to the way that the fundamental nature of the gravitational force is signaled by the fact that it depends solely on a body's mass rather than its chemistry. "Since I had always regarded the search for the absolute as the loftiest goal of all scientific activity," Planck wrote, "I eagerly set to work."[4]

In an often-told tale—which we relate in condensed form in the interlude—Planck discovered that he could produce a formula for the Reichanstalt data if he assumed that materials absorb and emit light selectively, only in integer multiples of a certain amount of energy that he called $h\nu$, where the constant h now bears Planck's name, and ν is the frequency of the radiation. If E is the energy, and n an integer, Planck's radiation formula is $E = nh\nu$. Planck later said he made this assumption out of "sheer desperation." It wasn't an explanation but a mathematical trick, he realized, and he thought eventually the idea could be discarded. For the moment, though, he was happy; the formula worked.

Few paid attention to Planck's idea. An exception, just five years later, was an unknown twenty-six-year-old patent office employee named Albert Einstein. In a famous, ultimately Nobel Prize–winning paper on the photoelectric effect, Einstein explained Planck's formula by the radical suggestion that light energy *itself* comes in multiples of h. Quanta of light energy would later become

The term "blackbody radiation" put to new uses, on the hull of a robotic sailboat built by students at the Franklin W. Olin College of Engineering.

known as photons. The quantum, in short, was not a mathematical trick, as Planck thought, but a physical reality. This, Einstein proudly wrote a friend, was "very revolutionary."[5]

Few took Einstein's idea seriously. But by 1910, physicists' interest in the quantum had kicked up a notch. One reason was that all attempts to eliminate the need for it, or reconcile it with classical physics, had failed. The quantum was cropping up all over, in molecular theory, heat conduction in solids, and elsewhere. Wherever scientists looked in the subatomic world, there it was, around every corner, energy in $h\nu$ multiples and $h\nu$ multiples only. It was like a guest whom you were obligated to invite to an event, even when you knew it would be awkward.

In 1911, a group of twenty-one leading European scientists

First Solvay Conference, Brussels, 1911, on the theory of the quantum. *Left to right, seated*: Walther Nernst, Marcel Brillouin, Ernest Solvay, Hendrik Lorentz, Emil Warburg, Jean Perrin, Wilhelm Wien, Marie Curie, Henri Poincaré. *Left to right, standing*: Robert Goldschmidt, Max Planck, Heinrich Rubens, Arnold Sommerfeld, Frederick Lindemann, Maurice de Broglie, Martin Knudsen, Friedrich Hasenohrl, G. Hostelet, E. Herzen, James Jeans, Ernest Rutherford, Heike Kamerlingh-Onnes, Albert Einstein, Paul Langevin.

gathered in Brussels for a summit conference about the quantum.[6] Underwritten by the Belgian industrialist Ernest Solvay, it was organized by the German physical chemist Walther Nernst. Nernst had initially rejected the quantum idea as "grotesque" but then admitted it was indispensable. Quantum theory has proven so useful, Nernst wrote to select participants, "that it is the duty of science to take it seriously and to subject it to careful investigation."[7] Planck was especially enthusiastic, having worried that not enough people were taking the idea seriously. As for himself, he wrote back, "for the past 10 years nothing in physics has so continuously stimulated, excited, and irritated me."[8] The lesson of the quantum, Planck told the gathering, was that classical physics "is obviously too narrow to account also for all those physical phenomena which are not directly accessible to our coarse senses."[9] In at least one area of the world, different rules apply from the usual Newtonian ones.

First International Attention

The 1911 Solvay conference—a quantum summit—was a turning point in the quantum's respectability, spreading the news beyond Germany to elsewhere in Europe. British scientist Ernest Rutherford brought word back to England, where he shared his excitement with an entranced young Danish visitor named Niels Bohr. The French scientist Henri Poincaré was so enthusiastic that he published a six-paged paper and gave lectures on the quantum before his death in 1912. In "The Quantum Hypothesis," he pointed out that the first person to see a simple collision—even an apple striking the ground—must have assumed it was a discontinuous event. Science, however, has shown that this is an illusion: what looks to be discontinuity is instead the result of rapid but continuous changes of forces encountering each other. Will something similar happen to the quantum, and we are simply not looking at it carefully enough? "Any attempt at present to give a judgment on these questions would be a waste of paper and ink."[10]

Americans were skeptical. The United States was then a backwater in physics; its physicists were mainly experimenters who viewed theory cautiously. They had resisted relativity theory, and were even more doubtful about the quantum.[11] These doubts were not mere pigheadedness; scientists had reason to be reluctant. "Classical physics exerted a strong appeal for physicists of every nationality by virtue of its own beautiful inner structure and the broad scope of the phenomena which it could explain," writes the physics historian Katherine Sopka. "In contrast, quantum theory at that time appeared confused and useful in only a relatively small number of instances."[12] Even the German quantum physicist Wolfgang Pauli told his comrade Werner Heisenberg a decade later, "it's much easier to find one's way if one isn't too familiar with the magnificent unity of classical physics."[13] American physicists who had been attracted to their calling by its elegance, beauty, predictability, and universality were struggling with an upstart theory that was—by contrast—ugly, weird, unpredictable, and apparently quite specialized. This passage from about 1915, evidently from a student presentation, aptly voices the typical American attitude toward the matter:

> I presume that all of us would agree that the quantum theory is quite distasteful. In working with the theory we have no definite mechanical picture to guide us nor have we any definite clear cut principle as a basis of operations—Physicists everywhere have been making strenuous efforts to find a method of escape from the theory. If such a method could be found I presume that we should all breathe a sigh of relief and sleep better thereafter.[14]

Robert A. Millikan (1868–1953), of the University of Chicago, was determined to help his colleagues sleep better. He, too, was

Robert Millikan (1868–1953).

scientifically conservative; he didn't believe Einstein's theory of relativity ten years after it was published, and continued to express a belief in ether, whose existence had been all but demolished by the Michelson-Morley experiment. He had spent six months in Europe in 1912, when he met Planck and attended his lectures, but remained utterly skeptical of quantum theory. Millikan was ambitious, and thought he could unhorse the false knight and kill the quantum idea the old-fashioned way: experimentally. He refurbished his laboratory with state-of-the-art equipment to test the predictions of Einstein's 1905 paper. Some of these predictions had already been tested and seemed to be positive, but Millikan was sure more precise measurements would tell a different story.

Millikan was shocked that his experiments confirmed each of Einstein's predictions. His astonishment and dismay are on display in his paper announcing the results. He consoled his readers by telling them that he'd heard that even Einstein didn't believe his own theory. In the end, however, he was forced to concede that Einstein's quantum theory "actually represents very accurately" the photoelectric effect.[15]

Meanwhile, the quantum kept cropping up in explanations of subatomic phenomena. In 1913–14, the young Danish physicist Niels Bohr, who had been interested in the quantum because his PhD thesis had shown that classical physics could not explain the electromagnetic properties of metals, applied it to the atom in a major breakthrough for atomic physics (see Chapter 3). Classical theory predicted that orbiting electrons would radiate energy and collapse into the nucleus. Bohr showed that if you assume electron orbits can only possess angular momentum in multiples of a new natural unit, $h/2\pi$, then the electrons do not have an infinite number of possible orbits about the nucleus, as planets do about the sun, but only a tiny selection—a result that dramatically dovetailed with experimental evidence. Another phenomenon discovered around the same time was the Stark effect, or the fact that spectral lines are split by an electric field. The effect was discovered in 1913 and explained by quantum theory in a 1916 paper that concluded, "It seems that the efficiency of the quantum theory borders on the miraculous and that it is by no means exhausted."[16] The quantum, a scientist observed, had turned into a "lusty infant."[17]

Millikan's experiment served as a turning point in the reception of quantum theory. In its aftermath, writes science historian Max Jammer, "the quantum of action became a physical reality, accessible directly to experiment, and Einstein's conjecture of light quanta was endowed with physical significance and an experimental foundation."[18] Still, it took almost another decade before light quanta were generally accepted.

Quantum of Solace

All of this discussion took place within the physics community, with little reaching the public. Ordinary language continued to use "quantum" in its traditional meaning of "amount," which could be large or small and even negligible, and was applied to all aspects of human life in expressions such as quantum of trade, quantum of naval strength, quantum of wealth needed for a good life, and so forth. Sometimes the term appeared in the Latin phrases *quantum meruit* (the amount that one merits) or *quantum sufficit* (as much as suffices).

In 1959, Ian Fleming published a short story in *Cosmopolitan* magazine entitled "Quantum of Solace," later reprinted in his collection *For Your Eyes Only*. It is one of the British author and former naval intelligence officer's finest and most moving stories. Not a spy thriller, it is a serious story that Fleming wrote while his marriage was failing. Late one night the main character, the Governor of Nassau, tells Bond a heart-wrenching tale about an acquaintance illustrating that human relationships can survive the worst disasters if both partners retain at least a certain finite amount of humanity. When partners stop caring, and "the Quantum of Solace stands at zero," the pain cannot only end the relationship but cause the partners to destroy each other.

Fleming's usage of the word "quantum" is close to Planck's—it means a finite amount of some quantity, enough to make a significant, structural difference. (The 2008 James Bond movie shares nothing but its title with the original story.) That meaning began to appear in public about 1916, when physicists, newly confident that the quantum was not a mathematical trick but physically real, began to speak more often about it.[19] The number and audiences for such talks soon grew as more news of the quantum reached the public. Planck was awarded the 1918 Nobel Prize "in recognition of the services he rendered to the advancement of Physics by his discovery of energy quanta." In 1919, Einstein became a household name when predictions of his theory of general relativity were confirmed. Popular accounts of Einstein and his theory often mentioned that this genius was also working on quantum theory, further elevating its status. Still more publicity came in 1922 when Einstein received his Nobel (for 1921), "for his services to Theoretical Physics, and especially for his discovery of the law of the photoelectric effect," that is, for his contributions to quantum theory. In 1923, Millikan received a Nobel "for his work on the elementary charge of electricity and on the photoelectric effect."

The press that appealed to educated audiences increasingly mentioned Planck's idea to describe a world that was not continuous in the Newtonian sense. From the 1929 *Manchester Guardian*:

"Nothing can happen in the world unless it performs according to a multiple of *h*. For instance, the energy with which you bite a piece of rump steak must be measured by an exact number of *h*'s. It cannot be so many *h*'s with a bit over, as it seemingly ought to be on a chance bite."[20]

If Planck's name was occasionally dropped in the 1920s and 1930s by serious journalists, in our time he has become a pop culture star, the poster child for the pixelated world, the first of those to detect cracks in the foundation of the Newtonian world. His formula $E = h\nu$ is one of the few equations recognizable by the public.[21] It is even the subject of a joke that won an award for physics humor by virtue of its economy. To understand the joke requires knowing simple mathematics, and the fact that ν is the Greek letter "nu." The joke then runs as follows:

Q. What's new?
A. E/h

Planck's equation, too, has become a celebrity, appearing on T-shirts, coffee mugs, and other places that often sport celebrity images. Like a celebrity, a large portion of the public recognizes it and thinks it important, but does not really know why. Like a celebrity, too, its public profile and influence have been enhanced by a social process; in this case, by popular science books and articles that have heralded its transformational role in the history of science.

Some journalists and writers, however, became convinced that Planck's work had not only opened up a new scientific era, but a new cultural one as well. At the end of 1930, for instance, the *New York Times* published an essay entitled "Science Needs the Poet." The Newtonian universe, with its rigorous "engineering conception," left no room for artists, except to "glorify mechanism," the author wrote. At last, "the machine that they worship has been thrown over by science," for "science in the form of mathematical physics waxes more idealistic with each new discovery." Once upon a time, Greek and Latin poets knew the science of their day. Our poets, the *Times* said, would do well to do likewise:

> Miracles? The poet will discover them in Planck's quantum theory, which explains radiation as if it were composed of bullets instead of waves. Quanta of light and electrons fly about with no respect for the old remorseless laws of cause and effect, and behave as if they were endowed with free will. Fairies are no more whimsical and Aladdin's lamp not half so wonderful. Matter is a mere wraith. An atom is something ghostly from which radiations emanate. The cosmos is no longer a machine which moves in a predictable way.[22]

The world is full of new possibilities for artists, poets, and philosophers, the *Times* continued. "What we need is a Lucretius who will imbibe at the spring of Einstein, Planck, Schroedinger and Heisenberg, compose a modern 'De Rerum Natura' and interpret the mystery and beauty that lie in and beyond the electron and space, curved and finite."

The quantum has become an even bigger celebrity in our era, and opened new expressive possibilities for artists and philosophers. It has done more than just making possible an updated version of *De Rerum Natura*.

Planck, of course, had none of this in mind; thoughts about the implications of his scientific concept for art and literature might have bemused and perhaps embarrassed this deeply conservative man. In 1947, the last year of his life, Planck was invited by the Royal Society in London to a Newton celebration. The event was held in a large hall, and distinguished guests were honored one by one, a page reading off their names and countries from a list in a loud voice. Planck had been specially invited and was not on the list. The page, confused, suddenly had to improvise and stumbled: "Professor Max Planck—from no country!" The attendees laughed good-naturedly and gave Planck a standing ovation as the nearly ninety-year-old man slowly and haltingly rose to accept the warm applause. Embarrassed, the Society added Planck to the citation list the next day, but Planck insisted that he be identified as "from the world of science."[23]

Max Planck Introduces the Quantum

> Never in the history of physics was there such an inconspicuous mathematical interpolation with such far-reaching physical and philosophical consequences.
>
> —Max Jammer, *The Conceptual Development of Quantum Mechanics*

The Newtonian Moment began to unravel in studies of the behavior of light.

At the end of the nineteenth century, the electric industry was booming. Telegraphs had existed for decades, but many new applications for electricity were in sight. In 1882, the American inventor Thomas Edison built an electric network able to provide 110 volts to a few dozen people in Manhattan, demonstrating in principle how electricity could be widely distributed. Two years later the Anglo-Irish engineer Charles Parsons built the first steam turbine and attached it to an electric generator, demonstrating in principle how huge amounts of electricity could be produced cheaply to supply networks.

Electric lamps were one of many sorts of devices that would run off those networks. The principle behind electric lamps is that all bodies emit heat and light at a range of frequencies, in the visible spectrum and beyond; night glasses and thermal imaging work by picking up this light. This range of frequencies has a particular profile. The hotter the temperature, the more the total intensity of the emitted radiation, and the higher the peak frequency. The intensity tapers off sharply at still higher frequencies, and more slowly at lower frequencies. An iron poker glows red because at that temperature most of its visible emitted radiation peaks in the red range. As its temperature rises, the peak becomes sharper and crests at still higher frequencies; the poker glows white instead of red. Each material radiates the same spectrum of colors at the same temperature *if* they all were perfect absorbers, i.e., perfectly black. In fact, many substances are pretty good absorbers, i.e., poor reflectors, so they make fairly good black bodies. Electric lamp manufacturers sought to build bulbs to glow with the maximum amount of white light while consuming the minimum amount of energy.

In the 1880s, many countries took steps to stabilize and foster their electric industry by creating laboratories to develop and supervise standards for electric lamps and other products. Scientists at Germany's Reichanstalt built special ovens to measure blackbody radiation and collect data on temperature and light emission at high frequencies. Reichanstalt theorist Wilhelm Wien (1864–1928) produced a formula to describe the profile of the electromagnetic radiation emitted at each temperature. It showed that the intensity of emitted light increases with temperature, but the increase is not distributed equally across all frequencies, and shifts or is "displaced" toward shorter wavelengths.

Planck was dissatisfied with Wien's law. Not because it didn't work, but because it seemed to be only an inspired guess. Why were frequency, intensity, and temperature related this way? Why did the constants have the particular values they did? The answers ought to be absolute, derivable from fundamental laws of electromagnetism and thermodynamics. Trying to find a way, Planck modeled the way materials absorb and emit radiation in the traditional way by picturing them as containing a set of "resonators," electric charges that oscillated back and forth as if attached to Hooke's Law–like springs of varying flexibilities. Their oscillation rates—their frequencies—depended on the stiffness, so to speak, of the springs. When the material absorbed energy, the resonators oscillated more energetically; when they shed energy, they oscillated less. A material could be treated as having an infinite set of such springs of different stiffnesses as it absorbed and emitted energy of different frequencies. In 1899, Planck was thrilled to find he could use this model to derive Wien's law.

His joy was short-lived. Early in 1900, Reichanstalt experimenters built more sensitive ovens to measure longer (infrared) wavelengths. On October 7, experimenter Heinrich Rubens informed Planck that, in this range, they were getting results slightly at variance with Wien's formula, with the spectral function more proportional to the temperature—more in line with the classical formula, later known as the Rayleigh-Jeans formula, than with Wien's.

Some scientists could live with the discrepancy. Planck had shown that Wien's law was fundamental; surely the discrepancy was experimental error, which does occur especially in new instruments and delicate measurements! But Planck knew Rubens and the other Reichenstalt researchers, trusted their results, and was skeptical that some unknown factor was throwing them off. He went back to work and produced a new formula, presenting it at the Berlin Physical Society on October 19, 1900. "[A]s far as I can see at the moment," he told his colleagues, this formula "fits the observational data published up to now as satisfactorily as the best equations put forward for the spectrum." He concluded, "I should therefore be permitted to draw your attention to this new formula which I consider to be the simplest possible."[24]

Rubens was at the Berlin Physical Society the night Planck presented his formula. Excited, Rubens returned to his lab to check the formula against his own measurements, and found they matched. The next morning, he visited Planck to tell him the news. A pair of other experimenters, who had reported results appearing to diverge from Planck's formula, soon found that they had made a mistake and that their results also matched the formula.

Planck was still not satisfied. In his eyes, his formula had the

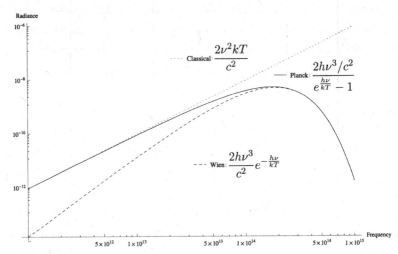

The three proposed laws for blackbody radiation given in radiance (or intensity) versus frequency. Wien's law, which follows the dashed line below, fit the experimental data for emitted radiation at high frequencies, describing—to the right of the graph—how it peaks at a specific frequency, which is higher the hotter the temperature. The classical formula, which follows the dotted line at the top, did not fit those data, but was assumed to describe lower frequency data to the left of the graph. Planck's formula, the solid line connecting them, is therefore like Goldilocks's ideal: neither too large nor too small. He initially assumed it to be simply a mathematical "trick," pending more research, that showed a way to connect the classical law at low frequency and the Wien law at high frequency.

same defects as Wien's. It was still an "empirical formula," a guess without "true physical meaning."[25] Again he set to work. In November 1900, he used his extensive knowledge of thermodynamics to calculate the way energy was distributed among the resonators in his model, hoping for some clue that would allow him to derive his formula from fundamental physics. "After a few weeks of the most strenuous work of my life," Planck later remarked in his Nobel Prize lecture, "the darkness lifted and an unexpected vista began to appear."[26] Trying to derive his equation he made what he later called "an act of desperation" because "a theoretical interpretation therefore *had* to be found at any cost, no matter how high."[27] He found he had to assume that the total energy E distributed among

N resonators consisted of a set of "energy elements" ε, whose value was related to a new fundamental constant of nature that he named simply *h*. It was ad hoc, invoked to apply only to the mechanism of the interaction between matter and the light it radiated.

For high enough frequencies, the term 1 in the denominator is negligible compared to the huge exponential term. Leaving out the 1, the exponential in the denominator becomes an exponential with negative exponent in the numerator, exactly reproducing Wien's formula. For small frequencies the exponential in the denominator approaches 1 and only the difference of the two terms is important, and the difference may be approximated as $h\nu/kT$, so that in the whole expression the power of ν is reduced by 1 and a factor kT appears in the numerator. This is precisely what classical theory—no quantum effects—would imply, though Planck may not have been aware of this in 1900. At least in hindsight, Planck's formula smoothly interpolates between classical physics at low frequencies and Wien's formula at high frequencies. "This interpolation," writes historian Max Jammer, "though mathematically a mere trifle, was one of the most significant and momentous contributions ever made in the history of physics."[28]

Planck presented this to the Berlin Academy on December 14, 1900. In this key passage, "Planck's constant" appears in print for the first time in history:

> We consider, however—this is the most essential point of the whole calculation—*E* to be composed of a well-defined number of equal parts and use thereto the constant of nature $h = 6.55 \times 10^{-27}$ erg sec. This constant multiplied by the common frequency ν of the resonators gives us the energy element ε in erg, and dividing *E* by ε we get the number *P* of energy elements which must be divided over the *N* resonators.[29]

This last sentence states in words what is now known as the Planck energy formula ε = *h*ν (or *nh*ν, some integer multiple of this

basic amount), one of the most fundamental equations that structure the world.

Neither in that passage, nor anywhere else in the paper, does
Planck refer to his idea as a "quantum." He uses the word once in
the paper, to refer to the "quantum of electricity e," or the basic
electric charge, but does not apply it to his mathematical trick.
He was not yet sure it was basic. Neither he, nor anyone else,
guessed the magnitude of the revolution in physics that was set
in motion. The passage amounts to saying—though Planck does
not say it explicitly—that the resonators of a certain frequency
could not have just any energy, as classical physics implied, but
that the allowed energies would be separated by integer multiples
of $h\nu$. Jammer writes, "At that time Planck apparently was not
yet quite sure whether his introduction of h was merely a mathematical device or whether it expressed a fundamental innovation
of profound physical significance."[30] Planck had indeed achieved
something fundamental, but not what he expected. He had acted
in a supremely conservative fashion, trying to save as much of the
existing theory as possible, introducing as little as possible of the
new in order to explain the experimental results that he trusted.[31]

Five years later, Albert Einstein turned mathematical trick into
physical concept. In 1905, he completed a PhD at the University
of Zürich, and published four extraordinary scientific papers that
broke with classical physics. One was on the photoelectric effect,
the fact that light shown on a metal makes electrons jump off its
surface. Classical theory predicted the electrons' energy should
depend on the light's intensity. It didn't; it depended on the light's
frequency—if you shine more intense light on a surface more electrons jump, but with the same energy. While Planck had said,
cautiously, that you had to introduce the funny h term into the
formula because the resonators could only absorb and emit light
with energies in multiples of $h\nu$, Einstein now said h was a property
of light *itself*. Light is grainy, localized in space and with energies in

multiples of $h\nu$, "quanta" of light, later called "photons." "Energy, during the propagation of a ray of light," he wrote, "is not continuously distributed over steadily increasing spaces, but it consists of a finite number of energy quanta localized at points in space, without dividing and capable of being absorbed or generated only as entities."[32] If so, the energy of the photons you shine on the surface of a metal have to be at least as great as the sum of the final energy of the electron that it causes to jump off the surface, plus the energy required to let the electron escape from the metal (the work function, or W: $E_{max} = h\nu - W$). It is like the way the energy of a bullet at the muzzle of a rifle is less than what it was when first exploded because of the energy lost traveling up the barrel.

For a while Planck struggled, in a way many of his colleagues (he once wrote) found almost tragic, to fit the idea of h into the conventional physics framework. In vain. As long as things involved high temperatures and long time periods, the existence of h could be ignored—but every time high frequencies or small temperatures were involved, it demanded entry back into the calculations. You *had* to have the quantum to produce Wien's formula for blackbody radiation at high frequencies, while classical physics could give you *only* what was soon called the Rayleigh-Jeans formula, which described such radiation at low frequencies but gave rise to an impossible prediction of infinite radiation called the "ultraviolet catastrophe" at the high end.

At the 1911 Solvay conference in Brussels, Planck pointed to yet another bizarre implication. His constant is not a specific amount of energy, but states a proportionality between two quantities. The equation $E = nh\nu$, for example, relates energy and frequency; if one goes up, so does the other. Physicists call constants of proportionality between energy and frequency "constants of action." At the conference, Planck discussed some startling implications of the fact that h is a unit of action via the notion of phase space, or the abstract representation of all possible key values that a system can have. The phase space of a moving object, for instance, can be represented in a graph—let's say the size of a tabletop—with position on one axis, and momentum on another. In classical Newtonian physics, every point on the table (of which there are an infinite number) is a possible position for an object. When an object changes its position and momentum, it moves as it were from one point to another—tracing out a line—and Newton's laws can be used to predict its path. Planck noted that, in the quantum realm, such an object does not have an infinite number of positions. The table is formed of patches—pixels, we'd say today—each of area h, representing a possible state—range of values of position and momentum—for the object. Its possibilities are now vastly restricted. It cannot move continually from one point to another, but is stuck in a region at least as big as one quantum. Each such pixel is a possible space of action, and the object's motion is discontinuous as it hops from one to the next. With the benefit of hindsight, many of the later principles of quantum theory, including the uncertainty principle, are implicit in this realization that phase space is pixelated or "quantized."

The quantum was interjecting warps and bubbles into more and more places of the universe, and opening a new era in science.

Chapter Three

Quantum Leaps

Quantum Leap is the name of a US comedy/adventure/science-fiction TV series of the early 1990s whose time-traveling protagonist—a holder of six doctorates, whose "special gift was quantum physics"—took a different jump through space-time each episode. In the 1980s, the British computer firm Sinclair launched a supposedly game-changing machine called the Sinclair QL (which they shortly abandoned). In 2000, a camp called "Quantum Leap Farm" was founded in Florida to help disabled equestrians change their lives through building new relationships with horses. In 2013, Bosch introduced "a quantum leap forward in sunshine" with Finish® Quantum® dishwasher fluid. In 2014, Royal Caribbean christened a cruise ship, *Quantum of the Seas*, that it said makes a "quantum leap forward in cruising" at 1,141 feet long and 167,800 gross tons; by comparison, the *Queen Mary 2* is a mere 1,132 feet long and 148,528 gross tons.

How did "quantum leap" leap from a scientific term that applied to subatomic state transitions to an idiom meaning "big jump"? The answer does much to explain the stories of other words and phrases that we discuss later in the book. It shows how it is possible for technical scientific language and imagery to become metaphors, for those metaphors to evolve in new ways, and then for them to take their place as a meaningful part of discourse in art and literature.

Niels Bohr

The notion of "quantum leap," if not that exact phrase, was introduced into physics in large measure by Niels Bohr's application of the quantum to atomic theory.

Niels Bohr, a legend of physics, was one of the most influential twentieth-century scientists. He had a long, oval, big-cheeked face and thick, wavy hair that he combed straight back from his forehead. Physically, he was imposing—seemingly aloof and stern—but his colleagues knew that he was kind and pensive. He also had a relentless, even bold, curiosity, as illustrated by a story told by the Russian-born American physicist George Gamow, a student of Bohr's. Bohr had a fondness for Hollywood Westerns that he shared with his students. He noted that while the villains of these movies always pull their guns first, the good guys still always manage to shoot first and kill the villains. Wondering why, Bohr theorized that this had psychological causes: drawing first, the villain has to think, which slows the action—but the good guy is acting automatically and reflexively, and therefore acts quicker. Gamow and his fellow students were dubious. The next day Gamow went to a toy store, bought a pair of Western-style cap guns and holsters, and insisted that Bohr experimentally test his theory. "We shot it out with Bohr," Gamow reports, "he playing the hero, and he 'killed' all his students."[1]

Science history books usually cite his main contribution to physics as "the Bohr atom," his use in 1912–14 of the still-recent quantum idea to solve problems raised by Rutherford's discovery that the positive electric charge in an atom is localized in a very small, massive "nucleus."[2] But his legendary status derives from the intensity with which he spoke to people, one-on-one. Bohr thought by talking aloud. He would entice a student, friend, or colleague

Niels Bohr (1885–1962).

into a long walk—and then instead of beginning a dialogue would begin a relentless exposition of his ideas, stating, reworking, and then restating them yet again. If his companions objected, he would push them repeatedly into rethinking their principles. If he sensed they were not agreeing, his voice grew stronger, and Bohr would not let up until he was sure they had capitulated. He browbeat Erwin Schrödinger into a (temporary) retraction of his ideas, while Werner Heisenberg once broke down in tears under Bohr's relentless questioning. Bohr's most challenging, and frustrating, conversational partner was Albert Einstein; the two carried on an extended debate about the fundaments of quantum theory until Einstein's death in 1955, as we discuss in a later chapter.

Bohr proposed his revolutionary model of the atom only two years after defending his dissertation on electron theory at the University of Copenhagen in 1911. Thanks to a fellowship from the Carlsberg Brewery—which financed bright Danish students—Bohr traveled to the Cavendish laboratory in Cambridge, England, whose chief was the experimenter J. J. Thomson, who had helped discover the electron. But Thomson did not find Bohr's research interesting, and was put off by the persistent Dane. At the end of 1911, in a visit to Manchester, Bohr met Ernest Rutherford, the world's foremost experimenter in radioactivity, and the two found they were a much better match. A few weeks later, Bohr decided to leave the Cavendish and join Rutherford's lab, moving to Manchester in March 1912. There Bohr laid the groundwork for a revolution in physics by incorporating the quantum into atomic structure. "His move," writes the historian John Heilbron, "symbolizes the end of one era, and the beginning of another."[3]

At the time, physicists could still only guess how the atom was structured. Thomson's work showed that atoms contained particles called electrons that were negatively charged. The existence of a proton—a particle to carry positive charge—had been proposed but not generally accepted. Rutherford argued that all positive charge in atoms resided in the central nucleus, and proposed that the electrons whirled about it sort of the way planets orbit the sun. But the idea attracted little attention, and was not promoted much even by Rutherford.[4] It is not mentioned in the proceedings of the 1911 Solvay conference—which Rutherford attended—and was only briefly mentioned in scientific journals and correspondence of the day.

Scientists had good reason to be skeptical. Electric charges are not like planets. When electric charges whirl about under the influence of another charge they radiate energy. The electrons in Rutherford's atoms would quickly shrink their orbits and spiral into the nucleus. According to the traditional laws of motion and attraction, Rutherford's idea would not work.

Bohr, however, sensed that the quantum played some bizarre role in making atoms stable. He was not the first to think so. "It was in the air to try to use Planck's ideas in connection with such things," Bohr once recalled.[5] But Bohr was in a special position to find the right way to solve the problem. He knew Rutherford's model in depth, and he was armed with the latest information about the quantum. The solution came to Bohr in several bold steps in 1912.

Why couldn't electrons orbit the nucleus any way they liked, similar to planets about the sun? Why didn't they radiate energy? The answer, Bohr felt, might have to do with the quantum of action—with the fact that phase space came in chunks. Action is related to energy multiplied by time. In the Newtonian world, action is continuously variable; the implication that it was atomic was bizarre and hard to picture. As one noted scientist wrote at the time, "an attempt to imagine a universe in which action is atomic leads the mind into a state of hopeless confusion."[6] Bohr's hunch was that, to the contrary, quantization of action might explain the very stability of the atom. To prove it, he would have to find a connection between the kinetic energy of the electrons whirling about inside the atom and their frequency. He was excited enough to send Rutherford a memo about the idea, though Bohr admitted it was "hopeless" to seek a "mechanical foundation," meaning a model. Bohr wrote his brother Harald, "Perhaps I have found out a little about the structure of atoms. Don't talk about it to anybody."

For a few months Bohr was stumped. Then a thunderbolt struck, having to do with spectral lines. These are the light given off by electrons inside atoms as they jump back and forth between orbits, absorbing and emitting energy in the form of light of specific frequencies in the process. Such lines had been studied for half a century, by scientists called spectroscopists. ("I'm not a very funny guy," runs an old spectroscopist joke, "I only know a few good lines.") In 1885, a Swiss schoolteacher named Johann Balmer made an astonishing find—the stuff of fairytales in science. Balmer, who had a background in math, managed to fit the frequencies of the hydrogen spectral lines to a formula that not only worked but predicted other lines of which he had been unaware. In modern notation, the formula producing the series of wavelengths λ is $1/\lambda = R(1/n_1^2 - 1/n_2^2)$, where R is a constant called the Rydberg constant, n_1 is an integer greater than zero, and n_2 is an integer greater than n_1.

"As soon as I saw Balmer's formula," Bohr said later, "the whole thing was immediately clear to me."[7] The formula showed a classically continuous quantity—wavelength—taking only values that are integer multiples of a basic unit. But that is a sign of quantization, Bohr realized, and the formula gave him a signpost for observing the quantum nature of atomic structure. When electrons absorb energy from photons, they "store" energy by moving to a higher orbit—and release it in the form of photons of a specific frequency in going back down. Easy math shows that the binding energy in the second orbit is a quarter the first, the third orbit a ninth, then a sixteenth, and so on. The energy lost in transition from a higher to a lower orbit is emitted in the form of light of a specific frequency: a spectral line. The quantum of action meant there are only certain allowable atomic orbits. It was as if there were only certain allowed orbits about the sun, and a planet could not make a smooth transition from one to another but instead would materialize at the allowable place. Bohr realized the Balmer series allowed him to connect the quantum of action and the spectral lines. It was a fingerprint of the atom's quantum structure.

Bohr worked feverishly for weeks. In March 1913 he mailed the first of a three-part article, "On the Constitution of Atoms and Molecules."[8] It showed how electrons do not have an infinite number of possible orbits about the nucleus, as planets do about the sun, but only a small selection. He ran into one serious difficulty. He could not explain why electrons do not fall back from their ground state into the nucleus. "Bohr circumvented this disaster by introducing one of the most audacious postulates ever seen in physics," wrote the physicist-historian Abraham Pais. "He simply declared that the ground state *is* stable, thereby contravening all knowledge about radiation available up till then!"[9]

Scientists now rethought Rutherford's strange atomic scheme— and also the quantum. One of the first influential accounts of Bohr's work appeared the next year: "Report on Radiation and the Quantum-Theory," a little book published by the British scientist James Jeans in London in 1914. It began by noting the recent discovery that Newtonian laws are not "fine-grained enough to supply the whole truth with regard to small-scale phenomena," and ended with Bohr's work. Jeans concluded: "The quantum-theory . . . represents a complete departure from the old Newtonian system of mechanics."[10]

Bohr's explanation worked beautifully—for hydrogen. It did not work well for other elements. When an atom has more than one electron, the interactions among the electrons serve to screen the positive charge of the nucleus, affecting how tightly bound the outer electrons are to it, introducing complexities that were not easily handled by Bohr's treatment. No matter: other physicists would soon step in to help.

Leaps, Jumps, and Jerks

Bohr did not use the words "leap" and "jump" in his early papers, and spoke simply of electron orbits or stationary states. However, his model said that electrons could have only specific energy values, meaning that the language of orbits and trajectories between orbits was unsatisfactory. Electrons could not transition smoothly between one atomic location and another—as a satellite changing orbits—but made the transition instantaneously somehow.[11] Within a few years, scientific texts and popularizations began to use more precise language and referred to "jumps," "leaps," and even—for a while—"jerks" between states.[12] The *New York Times* science journalist Waldemar Kaempffert remarked that "Energy is therefore discontinuous and atomic. It comes in jerks, like motion-pictures, but the jerks follow one another so rapidly that we see continuity."[13] Likening it to what happens in motion pictures was a wonderful image, extending an invitation to treat quantum jumps as a proper feature of our world.

The word "quantum" quickly became a metaphor for discontinuity. In 1929, the *Sun*, a US newspaper, noted that modern life had become governed by things that click. "Clocks, obviously," it wrote. "But also typewriters, adding machines, cash registers, speedometers, tachometers, stock tickers, automatic telephones, telegraph instruments—the whole tribe of appliances that operate by jerks are the masters of men who work. It is the reign of the *quantum* theory in industry."[14] Discontinuities are not necessarily random or uncaused; the jerks of all these devices are the products of continuously acting Newtonian forces. Still, the *Sun*'s metaphor captured a recognizable feature of modern life.

The *Observer*, another US newspaper, quoted a comment from an official of the licensing commission that there was no degree of drunkenness; the official assumed, evidently, that drunkenness was its own state that one was either in or not. "So does the new doctrine of the discontinuity of nature overflow into the sphere of the mind, and man threatens to behave like a quantum."[15] "Quantum" was now a name for things that came in on-off states rather than degrees.

Language has a "moment" of its own—in science, "moment" is a technical term referring to a tendency to twist things—and can often transform what words mean in unanticipated ways. This

From the webcomic *Saturday Morning Breakfast Cereal* (SMBC) by Zachary Alexander Weinersmith.

happened to the phrase "quantum leap." As life at the beginning of the twentieth century became more and more mechanized, and encountered larger and larger discontinuous transitions—in such things as populations and military might—the term "quantum leap" had a vitality and glamour. It was soon applied to any large, qualitative increase, whether of effort, money, or military might, with or without possible intervening steps. The first entry for "quantum leap" in the *Oxford English Dictionary* refers readers to a physics definition; the second, for nonscientists, defines quantum leap as "a sudden, significant, or very evident (usually large) increase or advance."

Whether you consider a quantum leap large or small depends on your scale. To paraphrase the astronaut Neil Armstrong, a quantum leap is a small step (in energy) for a man but a large, discontinuous step for an electron (see Interlude). Still, it is remarkable that "quantum leap" has come to be applied to changes on the human scale that are not only huge but not necessarily discontinuous (such as new dishwashing fluids and cruise ships).

Symbols Grow

Words are promiscuous; they do not have a fixed meaning. Their fluidity is limited, and they can be exploited and misused, but still they can generate new meanings. The change in meaning of "quantum leap" is not unique. Several words of ordinary language—including moment, force, gravity, and numerous others—have ended up as technical scientific terms, while several scientific terms—including complementarity, uncertainty principle, entropy, and catalyst—are often applied to ordinary life.

Words generate new meanings usually through metaphors. Metaphors contain two terms, a primary and a secondary. In "love is a rose," love is the primary term whose meaning is being explored,

"*I blame entropy.*"

while rose is the secondary term, used to elucidate the first. This is a "filtrative" metaphor, for it asks us to filter our perceptions of the primary term in light of certain well-known features of the secondary (love, like roses, is pretty yet thorny). The terms are not confused. A rose is not love; it remains in the garden, its identity unaffected. However, a new meaning has appeared—love's rose-likeness—that lets us understand our experience better.

Metaphors are particularly valuable when part of our experience is enigmatic—when the "correct" words are insufficient and we need new ones even if technically incorrect. The urge to metaphorize, what we'll call metaphorical compulsion, is a discontent: we feel the need to put words to our experience, the existing words don't fit, and we search for other words, any words, that help. Metaphors are in many ways like Rorschach tests; they are meaningful not because the primary term is really related to the secondary, but because of a shift that the secondary term works on our perceptions. Metaphorizing is a process by which we recognize in the act that something has improved our understanding—that something new has emerged.

For example, in "Here Come the Maples," a 1976 short story by John Updike, the protagonist Richard Maple ruminates about his decaying marriage when his thoughts are momentarily interrupted by a chance reading about subatomic discoveries (the italics are Updike's):

> He . . . read, *The theory that the strong force becomes stronger as the quarks are pulled apart is somewhat speculative; but its complement, the idea that the force gets weaker as the quarks are pushed closer to each other, is better established.* Yes, he thought, that had happened. In life there are four forces: love, habit, time, and boredom. Love and habit at short range are immensely powerful, but time, lacking a minus charge, accumulates inexorably, and with its brother boredom levels all.

Maple is trying to fill in what he intuits but cannot say. He does not think his marriage *is* subatomic physics. Still, in this passage, he finds physics terms useful in understanding the dynamics of his marriage. He is confused, wants to understand, and uses the best tools he has at the moment—the words of an article he happened to have stuffed in his pocket. The article could have been about almost anything—economics, sports, theater—and he would have seized on those rather than physics. Metaphors are pointers, and what matters is the phenomenon to which the metaphor is pointing— here, Maple's marriage—not what is being used to point.

Expressions, however, can stop being metaphors when we forget their origins, and cease to connect the expressions with the world from which they came. Think of what Americans call the "hood" of a car, and the English a "bonnet," which no longer prompts natives of either country to think of garments. Pointers, that is, can turn into names that lose track of their origins. Language is full of them, and linguists know numerous delightful examples. Our favorites include "peculiar," which stems from an ancient Greek word for cattle, and "bistro," which comes from the Russian word for "quickly," via Russian soldiers stationed in Paris after the defeat of Napoleon who loudly called for French chefs to deliver food with more haste.

This is one way laboratory words can be meaningfully used

in the human world. Only certain words make this discontinuous jump to becoming pointers and names. It tends to happen to scientific concepts that, the scientist Arthur Eddington wickedly noted, are "simple enough to be misunderstood." More charitably, let's rephrase this as "simple enough to be suggestive." Quantum concepts often met this criterion.

But scientific words can make this transition in other ways besides filtrative metaphors. In *creative* metaphors the priority of the terms is swapped. In an extraordinary linguistic reversal, the secondary term deepens in meaning through the metaphor to subsume its previous meaning and that of the primary term as well. The pointer becomes the pointed at.

In physics, for instance, a "wave" originally meant something that took place in a medium. However, its metaphorical extension to light (which does not require a medium in which to move) and then to quantum phenomena (which involve probabilities) changed its meaning. A "wave" is now not just a metaphor but the correct term for light itself, as well as other things such as probability variations that it did not originally name.

The phrases "quantum leap" and "quantum jump" represent the earliest, and most enduring, popular invocations of the quantum. The literal-minded may declare that these words are being used incorrectly at best, and emptily and manipulatively at worst. But not necessarily, any more than someone referring to a car's "hood" or "bonnet" speaks incorrectly. The way that "quantum leap" and "quantum jump" spread from the subatomic realm to everyday language provides the model for the path followed by other language and imagery discussed in this book. This process creates the opportunity for scientific language and imagery to appear, meaningfully, in art, literature, and philosophy.

Niels Bohr Uses Quantum Leaps
to Make Atoms Go

Perhaps I have found out a little about the structure of atoms. Don't talk about it to anybody.

—Niels Bohr to Harald Bohr, 1912

When Bohr introduced the quantum into atomic structure, his idea was extremely simple. His first paper, published in July of 1913, applied it to the simplest case, that of a hydrogen atom, consisting of a single electron orbiting a single proton. Bohr made several other assumptions for the sake of simplicity, such as that the mass of the electron is tiny compared to that of the nucleus, and that the electron is traveling at much less than the speed of light so that relativity does not come into play. He assumed the orbits were circular. He assumed that the electron did not radiate energy as called for by classical Maxwellian laws.

The mechanical description of such an orbit, in classical terms, set the centripetal force required to keep the electron going in a circular orbit (L^2/mr^3) equal to the electric force of attraction between the electron and the nucleus (e^2/r^2). That is,

$$\frac{L^2}{mr^3} = \frac{e^2}{r^2}$$

This gives rise to something similar to Kepler's laws for planetary motions. In this atomic picture, the nucleus is like the sun and the electron is a planet. If this were completely true, the electron could orbit the nucleus in an infinite number of possible ways, but would keep losing energy until it plunged into the nucleus.

Bohr then made another assumption—that only certain of these orbits are possible—and he used the idea of the quantum to pick out these select orbits. Allowed orbits, he said, are those for which the angular momentum of the electron is equal to an integer multiple of a constant formed by Planck's constant: $h/2\pi$, called "hbar" and represented as $\hbar$. Or that 2π times the angular momentum equals nh, with n an integer. The radius of the allowed orbits then follows:

$$r = n^2 \hbar^2 / e^2$$

When the electrons jumped or leapt from one of these orbits to another, the difference between their energies was:

$$\left(\frac{e^4 m}{2\hbar^2}\right) \left(\frac{1}{n_1^2} - \frac{1}{n_2^2}\right)$$

This led directly, through Planck's relation $E = h\nu$, to the following wavelength formula:

$$\lambda = \frac{2hc\hbar^2(n_2^2 - n_1^2)}{e^4 m n_1^2 n_2^2}$$

For $n_1 = 2$, this corresponds to the original "Balmer" series for hydrogen, with lines mostly in the visible spectrum.

The Bohr atom was a hybrid; it took the classical model and imposed a quantum constraint. It broke down almost immediately, because it failed to apply to anything more than hydrogen. That had to await further work by Bohr and others. But it was a momentous step, for it showed that the quantum was written into the building blocks of matter. Matter holds together because of h; though h is tiny, were it zero atoms would collapse in fractions of a second. The quantum leaps and jumps that h forces electrons to take make atoms stable and the microworld work. Why are all hydrogen atoms—or any atom—alike? Quantization provides the answer: because they all have the same structure! There is essentially only one way you can put together an electron and proton to make a stable hydrogen atom. The same is true for other atoms.

A quantum leap, we said, is a small step for a man, but a large step for an electron. How large? Let's consider the energy and distances involved. If we could leap 20 feet in the air, that would be huge, no? Now let's see if we can compare it to the leap in energy of an electron going from the first state in a hydrogen atom to the second. How far, let us say, would an electron shoot up in the earth's gravitational field at the earth's surface if kicked by the same energy? The answer is around a hundred million miles, meaning comparable to the distance from the earth to the sun! The comparison shows that the electron is making a truly gigantic leap— *not* a tiny one.

The electron jump energy is very small compared to essentially any energy involving a macroscopic object like a human jumper. But because the mass of an electron is so tiny this energy has a superenormous effect on the speed of the electron, involving accelerations that would crush a human being. The perspective of large or small for the energy change of the electron depends on what you compare it to.

Chapter Four

Randomness

Quantum Cloud by Antony Gormley, 2000. Galvanized
steel 29 × 16 × 10 m, permanent installation,
River Thames, Greenwich Peninsula, London, UK.

Its name is *Quantum Cloud*. It is a sculpture that rises 30 meters above a platform on the banks of the Thames, and from a distance looks like a huge pile of steel wool. Framed by the gray London sky, it does look like a cloud. Visitors to south London can't miss it when visiting the Millennium Park, or taking a river cruise. As you draw closer, you can make out the hazy shape of a human being in the center. Like the other sculptures in the series, this sculpture is made from steel rods by the British artist Antony Gormley. Entitled "Quantum Clouds and Domains" the series explores how form can arise out of seemingly random behavior.

Randomness figures differently in *Quantum Sheep*, the brainchild of Valerie Laws, a writer who lives in the north of England. In 2002 Laws spray-painted words onto the fleeces of sheep from a nearby farm. As the flock of about a dozen animals milled about, the words rearranged themselves and a new "poem" was created every time the sheep came to rest. A spokesperson for Northern Arts, which provided £2000 of funding for the project, described the result as "an exciting fusion of poetry and quantum physics."[1] Here is one of the resulting "haik-ewes":

> Clouds graze the sky
> Below, sheep drift gentle
> Over fields, soft mirrors
> Warm white snow

Laws explained to the BBC why she felt the project worth pursuing. "Randomness and uncertainty is at the centre of how the universe is put together, and is quite difficult for us as humans who rely on order," she said. "So I decided to explore randomness and some of the principles of quantum mechanics, through poetry, using the medium of sheep."

Some of her haik-ewes have been included in poetry anthologies. Plenty more are available: Laws figures that her method may produce 87 billion combinations. If read continuously at the swift pace of one every five seconds or so, it would take nearly a hundred thousand years to get through them all.

A different vision of quantum randomness appears in a book of poetry entitled *Quantum Chaos: Learning to Live with Cosmic Confusion*, by the author Tony Zurlo, who writes books for children. One of Zurlo's poems contains the following two stanzas:

> Could the dance of
> subatomic matter be
> improvisational?
>
> Can Reality
> be determined by a
> probability curve?

Cover of *Quantum Chaos*,
poetry by Tony Zurlo.

In the classical Newtonian world the answer to such questions is No! Matter doesn't dance; it is law-governed through and through, down to its tiniest pieces. Its largescale behaviors follow causally from its smallscale behaviors. A Laplacian intellect—the perfect mind armed with all relevant information envisioned by the Marquis de Laplace—would know every structure and could predict every event, the sum total of which comprises reality.

But we humans are not in that happy situation. We simply don't know how most of the world operates. Card games, coin tosses, lotteries, and dice rolls seem random, and our measurements with crude instruments are uncertain. To cope with these situations, scientists create theories of probability and measurement error. Blaise Pascal and Pierre Fermat, two French mathematicians who lived during Newton's time, developed probability theory for games, lotteries, and other contexts involving inanimate objects. The nineteenth-century Belgian mathematician Adolphe Quetelet applied statistics and probability to crime rates and other issues examined by social scientists. The French mathematician and philosopher René Descartes, among others, studied measurement error and variation.

Some key distinctions: *Random* events lack specific causes, reasons, or purposes. *Statistics* is the analysis of the frequency of events that have already happened, and patterns in them. *Probability* is the use of statistics to predict future events. *Uncertainty* is the amount of possible deviation of an outcome from a specific anticipated value. All these ideas were present in the Newtonian world, as ways of describing and addressing situations involving incomplete knowledge. Such tools were necessary in circumstances when you "could know the blueprint," writes historian of science David Lindley, "but not the shape and color of every brick."[2]

Today, randomness, uncertainty, and the inability to predict are often associated with quantum behavior, and viewed as permanent features of our world. Human figures arising from random rods, poems compiled by wandering sheep, and reality from probability are some instances where form arising from randomness is called "quantum." How did this happen?

Subversive Elements

The transition from classical to quantum physics was gradual rather than abrupt, because the ever-greater scientific use of statistics and probabilities acculturated scientists to a world ruled by statistics. This began in the nineteenth century, when several phenomena appeared in the Newtonian world that looked harmless but eventually turned subversive.

A first came via thermodynamics. The word was coined in the mid-nineteenth century to refer to the science of heat conceived not as a single substance, as it had been, but as a phenomenon arising from motions of numerous particles. In a reversal of the usual flow of methods from one discipline to another, physical science had to borrow social science tools to explain these motions. Thanks to Quetelet's work, statistics had become a well-developed resource to analyze birth and death rates, disease, crime, and so forth, where tracking each individual element was impossible. Scientists studying thermodynamics and other situations involving many motions realized that statistics might help them as well.[3] Studying gases was like studying society. Furthermore, this statistical information was exactly what scientists who studied complex systems wanted. Sociologists studying societies and scientists studying gases, say, were interested in the overall behavior, not the behavior of each piece. Let's take a science example: suppose you were studying how water boiled; would you really care when a specific bubble of a specific size appeared in a specific place, or would you instead want to know simply when bubbles formed?[4]

A second subversive phenomenon was Brownian motion, referring to the mysterious fact that, under a microscope, pollen grains seemed to jiggle and bounce around randomly. This unpredictable motion of inanimate objects could not be explained at the time. Yet a third was radiation, the phenomenon that certain kinds of atoms occasionally transform into other kinds, emitting energy in the process. First noticed in 1896 in uranium, it was soon detected in other elements. Radiation seemed spontaneous, with no particular cause. In 1900, studying the radiation of thorium oxide, Ernest Rutherford came up with a quantitative law. The intensity of the radiation, he said, fell off with time (what we now call its half-life) in a way consistent with each atom having a fixed probability of transformation per unit of time. This kind of law proved to work for every other radioactive element. For the moment, at least, Rutherford and others assumed this quantitative radiation law had a deeper, causal explanation and was another probabilistic shortcut. Rutherford's quantitative law resembled a mortality rate table—it was a mortality rate table for atoms. More information, he and others felt, would help scientists find a mechanism that would specify a causal explanation for each atomic death.

This accumulation of examples gave physicists an uneasy feeling that something mysterious lurked in small distance scales. That "something" was about to burst into the open.

At the end of the nineteenth century, the American philosopher-scientist Charles S. Peirce proposed that randomness lay at nature's very core; this, he felt, would prevent scientists from coming up with one final set of laws valid for all scales and disciplines, and require them to keep producing new laws. But as the twentieth century began, this was still idle metaphysical speculation, and looked like philosophical meddling. Even the subversive elements just mentioned seemed only to force scientists to use probability and statistics as the best available tools for understanding elements of nature that were merely out of reach.

The person who transformed statistics and probability from a convenient, though increasingly indispensable, tool into a structural element of the world was Albert Einstein, in a series of three papers on quantum theory written in 1916–17. These papers connected the wavelike properties of light, thermodynamic issues, and Rutherford's quantitative radiation law, to install probability at the heart of atomic behavior. The moment Einstein made this connection, he regretted it.

Planck's original conception of the quantum, in 1900, did not involve probability. Nor did Einstein's 1905 paper explaining the photoelectric effect.[5] After 1905, Einstein closely followed developments in quantum theory, though he was preoccupied with other matters, especially his development of general relativity. He left the patent office, and eventually became a professor in Berlin. Einstein was impressed by Bohr's three papers of 1913–14, which described atoms as storing energy absorbed from radiation by moving electrons from lower to higher states, and then as releasing that radiation by moving the electrons back down to a lower state; in the process, they emitted light at specific frequencies corresponding to the difference in energy of the two states. This is what creates the spectrum of an element, the "fingerprint" of its quantum structure. But even Bohr's work did not use probabilities in its discussion of the timing of electrons in their jumps between states. Only at the end of 1915, when Einstein completed his theory of general relativity, did he turn back to examining light quanta, and how they were emitted and absorbed by matter.

His efforts immediately panned out. In August 1916 Einstein

wrote to his close friend Michele Besso, "A brilliant idea dawned on me about radiation absorption and emission; it will interest you. An astonishingly simple derivation, I should say, *the* derivation of Planck's formula."[6] A few weeks later he explained to Besso why: Planck had simply stipulated that these oscillators work in that particular way, giving that particular relation; "Planck's papers provide no correlation between h and ε."[7] Einstein had adopted the strategy he had used to develop special and general relativity, and ignored the question of what kind of material emitted and absorbed the light to focus on a general derivation. But to do so required him to introduce a probability factor as a fundamental part.[8]

Einstein's papers of 1916 and 1917 were about a small corner of physics: atomic transitions. Still, these papers are milestones because they wrote probability into the fundaments of nature for the first time—not just as a short-cut—and in a way that would be vastly expanded by physicists to follow in the next ten years. Even then, the growing role of probability in science remained a technical feature known only to physicists.

All this would change after the quantum mechanics revolution of 1925–27, with the Schrödinger wave function and the uncertainty principle. References to quantum randomness and probability increased sharply after 1927, when the British astronomer Arthur Eddington, in *The Nature of the Physical World*, drew public attention to the role of randomness in general, and of the uncertainty principle in particular, in quantum theory. It used to be, he says, that physics came down on the side of predestination; now quantum theory has removed this. "Quantum theory is a determinism of probabilities, an iron formula, but a limited determinism is possible. Atoms parade, as unaccountable lunatics follow individually the determined general rules of the asylum." Thanks to Eddington and his popular book, the public became acquainted with the idea that quantum mechanics spelled the end of determinism as a philosophy, and the role of randomness in the universe.

On the eve of the 1930s, the *New York Times* editorialized: "That the '30s will be more sedate than the '20s because we shall be less self-conscious is one of the probabilities. But prophecy is extra-hazardous in view of what the relativity and quantum science of the late '20s has already done to probabilities and possibilities in nature."[9]

The Forman Thesis

In the first chapter we mentioned the externalist approach to the history of science, exemplified by Boris Hessen's claim that Newtonian mechanics was driven by the social forces of his time. Another classic example involves the history of quantum mechanics, and crystallizes around the work of science historian Paul Forman. Forman's influential and controversial work suggests that the Quantum Moment is not a proper moment at all, but a mere by-product of the crisis-obsessed, romantic, and antirational currents of thought that prevailed in post–World War I Germany during the Weimar Republic, named for the place where a convention wrote and adopted a new constitution for the recently defeated nation. His work adopts a particular approach to the question broached at the end of the second chapter: What is the relation between scientific activity and the cultural climate in which it emerged? In that chapter, we merely pointed out that novelists found quantum language and imagery especially appropriate to describing contemporary experience. Forman's approach is much different: he claims that the cultural climate actually caused the scientific activity of the day to adopt certain of its most distinctive features.

At the beginning of the 1960s, Forman was a graduate student in the history of science who participated in a project called "Archive for the History of Quantum Physics," a systematic attempt to collect documents, compile manuscripts, and conduct oral interviews to create a database on the early history of quantum theory. The ambitious project succeeded beyond expectation, and became "the main foundation for practically every historian working in the field since."[10] Like many other historians of the 1960s, Forman was inspired by the fervor of contemporary social and political events—he was working at the University of California, Berkeley, during the years of the free speech movement—to ask new kinds of questions about the role of social and political factors on the field he was studying. His doctoral dissertation, entitled "The Environment and Practice of Atomic Physics in Weimar Germany: A Study in the History of Science" and completed in 1967, examined the economic and political factors affecting German physicists immediately after World War I.[11] His colleague John Heilbron, also a historian, found Forman's dissertation a "revelation."[12]

Two years later, Forman published a contentious article on the discovery of X-ray crystallography, in which he attacked history that relies too much on scientific literature and the stories that scientists themselves tell about their work. The article was prompted by a volume of recollections of the discovery published on its fiftieth anniversary, edited by P. P. Ewald, who had been around at the discovery, entitled *Fifty Years of X-ray Diffraction* (1962). Forman's 1969 article on the discovery appeared in a history of science journal, and argued that histories that rely too much on papers and recollections of participants tend to reinforce ideologically based myths, overlook the messiness of discovery, and portray science as being purer in concept and execution than it is.[13] Ewald then blasted Forman's article as "worthless except as an example to which conclusions the study of literature under a biased point of view can lead." The exchange inspired in turn the journal's editor to remark that "physicists are as pig-headed and superstitious as any monk of the dark ages." The row was a classic illustration of the internalist tendency to take guidance from memoirs and recollections of participants, and the externalist tendency to view science history as the province of professional historians with their own set of tools. The conflict—which swirls around the question of who speaks most authoritatively, participants or trained outside observers?—is not easily resolved.

Two years after that, Forman published his most controversial paper, "Weimar Culture, Causality, and Quantum Theory, 1918–1927: Adaptation by German Physicists and Mathematicians to a Hostile Intellectual Environment."[14] The article argued, in what has come to be known as the Forman thesis, that the creation and adoption of acausal, probabilistic quantum mechanics sprang not so much from scientific developments but rather from the prevailing "currents of thought" characteristic of the culture of Weimar Germany. Forman claimed that these currents of thought affected the passing comments and attractions of physicists to the theory, yes, but more important the substance of the theory itself.

Forman begins by scorning historians who use "vague and equivocal expressions" in speaking of how culture influences intellectual activity, such as that culture "prepared the intellectual climate for" this or that development. The historian, Forman writes, must insist on a "causal analysis, showing the circumstances under which, and the interactions through which, scientific men are swept up by intellectual currents." Applied to quantum physics, such an analysis reveals that German scientists reacted to a hostile Weimar intellectual milieu by seeking not merely to burnish their image but "to reconstruct the foundations of their science" in a way that would "dispense with causality, with rigorous determinism."

Forman's article proceeds in stages. In the first, he sketches out how Germany's stunning and unexpected defeat in 1918 dramatically transformed the prevailing intellectual climate. The result was "a neo-romantic, existentialist 'philosophy of life,' reveling in crises and characterized by antagonism toward analytical rationality generally and toward the exact sciences and their technical applications particularly." The author who most epitomizes this current, according to Forman, was the German historian Oswald Spengler, who published, in 1918, the notorious *The Decline of the West*, which argued that civilizations are like organisms in that they follow a developmental pattern of rise, maturation, and decay. A key part of Spengler's argument was that all cultural activities— including science and mathematics as much as art, philosophy, and religion—are shaped by the character of the civilization in which they develop and have no meaning outside that civilization. Even the idea of causality, Spengler wrote, was a mere artificial construction of cultures as an ultimately hopeless attempt to stave off the irrational forces of fate. Forman cites a prominent minister of culture and education as saying, in a widely read essay of 1919, that "We must acquire again reverence for the irrational."

Physicists, who until then enjoyed vast cultural prestige, suddenly found themselves under attack. "Implicitly or explicitly, the scientist was the whipping boy of the incessant exhortations to spiritual renewal, while the concept—or the mere word—'causality' symbolized all that was odious in the scientific enterprise." Forman cites Planck as expressing the prevailing view of scientists that "the Weimar intellectual environment was fundamentally and explicitly antagonistic to science." Physicists adapted to the pressures of that climate by experiencing the presence of causality in their theories as undesirable, and then by rejecting it entirely.

Forman then describes what he calls "quasi-religious conversions to acausality" among physicists from 1919 to 1925. "As if swept up in a great awakening, one physicist after the other strode before a general academic audience to renounce the satanic doctrine of causality and to proclaim the glad tidings that the physicists are about to release the world from bondage to it." What we see, Forman writes, is "a capitulation to those intellectual currents in the German academic world" that he outlined in the first part of the article. That capitulation produced "the rise of a will to believe that causality does not obtain at the atomic level before the invention of an acausal quantum mechanics." Forman stresses "how conscious the physicists were of the fact that they were playing before an audience hostile to causality = mechanism = rationalism, and how anxious many were to play up to that audience."

It is no coincidence, Forman argues, that many German physicists began "dispensing with causality" suddenly in 1918, *"before it was 'justified' by the advent of a fundamentally acausal quantum mechanics."* The "inescapable" conclusion is "that substantive problems in atomic physics played only a secondary role in the genesis of this acausal persuasion, that the most important factor was the social-intellectual pressure exerted upon the physicists as members of the German academic community." The editors of a book about the Forman thesis write:

> As a methodological model, Forman's study has arguably been the most influential article ever published in the historical studies of science, with the possible exception of Boris Hessen's equally famous and controversial 1931 analysis of classical mechanics in *The Social and Economic Roots of Newton's Principia.* . . . [B]oth upended, each in its own way, the essentially Platonic ideology of science as a pure intellectual activity, a noble search for abstract truth supposedly in control of its intrinsic scientific method and of the criteria of true knowledge. Instead, both approached science as an essentially human, and thus also earthly, social and cultural activity, and accepted the necessary epistemological consequences of such an assumption.[15]

Forman's thesis suggests that the influence between science and culture is the reverse of the way we have been considering it. Instead of looking at the process of how science supplies cultural activity with new language and imagery, we should be looking at a deeper process: how cultural activity engages and transforms science.

Yet there are serious flaws in Forman's approach. Some scholars note that Forman misrepresents several of his important examples. Others point out that the earliest introduction of probability into quantum theory—Einstein's of 1916–17—precedes Forman's key date of 1918. Others attack Forman's evidence, noting that most of Forman's quotations from scientists come not from papers but from addresses to general audiences, and that he ignores the internal developments of physicists and their addresses to professional colleagues.

More importantly for us, we saw in the last chapter that the metaphorical transformation of the phrases "quantum leap" and "quantum jump" suggests that it would be starting off on the wrong foot to try to understand the connection between science and culture as Forman does, as one of causal interaction. The place to begin is with our own experience of the world. This is not neutral, it has a certain character, and this character encourages us to speak about the world in certain ways and discourages us to speak about it in others. This process, we saw, drives much of the appropriation of the language and imagery of science by art, literature, and philosophy. It is what motivated Antony Gormley to call his sculpture rising above the Thames *Quantum Cloud*, and Valerie Laws to call her poetry-producing process *Quantum Sheep*; it inspired Tony Zurlo to call his book of poems *Quantum Chaos: Learning to Live with Cosmic Confusion*.

Albert Einstein Shows
How God Plays Dice

Each of Einstein's contributions to quantum physics illustrates that he was the twentieth century's leading master of thermodynamics and statistical mechanics, Newtonian fields in which probability, randomness, and the theory of statistics rule. It is only fitting, then, that he was the one to introduce these into the core of quantum theory and write them into the foundations of nature itself—despite his later protestations that he did not think that God "plays dice."

In the papers of 1916–17 Einstein treated light as particles—the idea he had introduced in his 1905 photoelectric effect paper—and proceeded by asking, "What are the conditions on the distribution of light particles for a system of molecules and photons (as we now call them) to be in thermal equilibrium at a given temperature?" There were already well-accepted principles to determine the distribution of probability of various energy excitations of each molecule. From this, and other general information about how light should behave, Einstein deduced precisely the distribution of light energy, which Planck had obtained in 1900 by perhaps more questionable means.

Einstein imagined a piece of matter—he called it a "molecule," though it could also have been an oscillator or atom—bathed in light radiation. How must it absorb and emit radiation? Suppose it is in equilibrium, meaning it absorbs and releases amounts of energy at the same rate. Its rate of emission in the presence of this radiation bath is enhanced compared to what it would be if no other photons were around. Einstein used the ideas of thermodynamic equilibrium to conclude that the rate at which an atom would be able to emit a photon in the presence of an external field is greater than if the atom were in empty space. The field's presence is like that of a bunch of kids in a playground who come along saying, "Come on! Join us!" The photons come out more easily. Einstein calculated how this would work; how the processes of emission and absorption are related to each other. There's a spontaneous part—related to just the atom and electron—and another part related to the other photons—the "kids." But when is the atom going to emit the light, and which way it is going to go? Both are unpredictable. He dealt with the randomness of the "when" in his first paper. In the second and third papers, published in 1917, he explained the randomness of the "which way."[16] To even mention the idea of trajectory, he had to assume that there was a probability factor. "The postulated statistical law of emission is nothing but Rutherford's law of radioactive decay," Einstein wrote in one of the three articles. Einstein proposed that "probabilities themselves might have to be regarded as fundamental, basic, physical properties of atomic systems." This allowed a derivation of Planck's law "in an amazingly simple and general way."[17]

But at this very moment, Einstein had his first regrets. "It is a weakness of the theory," he wrote in the papers, "that it leaves time and direction of elementary processes to chance." There was no way of knowing when and in which direction the photon would take off. "Nevertheless," he concluded, "I have full confidence in the route which has been taken."

It's hard to know how seriously to take this remark, when Einstein's physics and his intuitions collided so directly. He clearly hoped that the route would take him somewhere other than the apparent direction in which it was heading. He wrote to Besso, "I feel that the real joke that the eternal inventor of enigmas [i.e., God] has presented us with has absolutely not been understood as yet."[18] The apparent abandonment of causality would gnaw at him from this time forward. As he would write to Max Born in 1920:

> The thing about causality plagues me very much too. Is the quantumlike absorption and emission of light ever conceivable in the sense of the condition of complete causality, or is a statistical residue left? I must admit that I lack the courage of conviction here. But only very reluctantly do I give up *complete* causality.[19]

A few years later, Einstein wrote to Born again that he could not believe that God plays dice. But Einstein himself was the first to imply that this was indeed how the deity operated, in his simple derivation of Planck's daring formula.[20]

Toward the end of his life, when talking to a group of surgeons in Cleveland in 1950, Einstein outlined one of the simplest examples of randomness in quantum physics. Suppose you have a wave coming to a boundary between two refractive media. Maxwell's theory guarantees that there will be a transmitted wave and a reflected wave. Make this wave so weak that only once a month a photon arrives: will it go through or bounce back? No way to tell! We can give the probability of both, but don't know which will happen. That is the clearest example of randomness in quantum physics. At the end of his talk, Einstein asked how it will further evolve. Perhaps, he says enigmatically yet honestly, the best way to answer is with a smile.[21]

Chapter Five

The Matter of Identity: A Quantum Shoe That Hasn't Dropped

Little boxes on the hillside,
Little boxes made of ticky tacky,
Little boxes on the hillside,
Little boxes all the same.
There's a green one and a pink one
And a blue one and a yellow one,
And they're all made out of ticky tacky . . .

And the boys go into business
And marry and raise a family
In boxes made of ticky tacky
And they all look just the same.
There's a green one and a pink one
And a blue one and a yellow one,
And they're all made out of ticky tacky
And they all look just the same.
                              —Malvina Reynolds, "Little Boxes"

Half a century ago, in an age that protested against conformity, the songwriter Malvina Reynolds expressed scorn in
an anthem of the 1960s entitled "Little Boxes," about houses that,
curiously, are different because they come in different colors, yet
nevertheless are "just the same."

Identity is an angst-stirring issue, and a social lightning rod.
Identity politics is political activity that focuses on advancing
group interests; gender identity relates to how individuals present
their masculinity and femininity; social identity refers to the por-

tion of one's selfhood derived from membership in various social gatherings; identity formation refers to the way individuals acquire a distinct personality; identity theft is a crime in which one person poses as another; and so forth.

Identity overlaps with, but is not the same thing as, individuality. Individuality is what makes each of us the distinctive person we are, while identity refers to our continuing sameness. Identity is also not the same thing as equality (except in mathematics), as the first "self-evident truth" of the American Declaration of Independence to the effect that "all men are created equal" illustrates.[1] That statement does not mean that all men are identical, but that each has an intrinsic value that is the same (though Kurt Vonnegut's dystopian short story "Harrison Bergeron," set in 2081 America where a "Handicapper General" who strictly enforces the Constitution insists that gifted people be handicapped if they stand out over others, explores the ramifications of equating equality with identity). The word *men*, of course, originally meant at least implicitly "white men," but was eventually broadened by constitutional amendments to include men of any race, and, later, women. In different circumstances we can adopt utterly different attitudes toward identity. Sometimes we insist on being treated the same way as everyone else, while at other times we insist on being recognized as special and unique, and scorn being treated as interchangeable.

Thus the concept of entities that both are identical in essence and identical in all detail can be threatening and disturbing. This is a feature of the microworld that is weird and has no counterpart in our direct experience. It raises a question that comes up over and over in quantum mechanics: How can literal identity in the microscopic realm give way to, and even create, the distinctness we see on the human scale?

In an elegant little book entitled *Seeing Double*, the physicist and historian Peter Pesic writes that the heart of quantum theory lies in its "radical innovations" on the question of identity and individuality.[2] Innovating concepts of identity and individuality is not easy. The Western literary and philosophical tradition, Pesic observes, has long explored the meanings, nuances, and dilemmas of identity and individuality—of what it means to be a unique being, a being with an interchangeable identity, and a being with a concealed identity—as early as the guises of Hector and Odysseus in Homer's *Iliad* and *Odyssey*, and in the way that gods sometimes concealed their identities to help humans, as Athena did to appear to Odysseus's son Telemachus and lead him to help his father return. For these ancient Greeks, Pesic writes,

> Identity is imprinted beyond human auspices, indelibly; the gods inhabit the larger domain in which individuality is forged and they can reveal its signs, though its mysteries lie in the dim realm of destiny. Here individuality is a *sacrament*: the union of something visible or perceptible—a name, a face—with something invisible and numinous: the hidden, indelible mark that sets Odysseus apart from all others, regardless of disguise or the change of time and age. It goes far beyond visible, distinguishable signs to a "primitive thisness," as later philosophers called it, meaning no particular quality (which might be changeable and could disappear), but an altogether different sense in which *this* person or thing is not *that*. Such deep individuality endures even in the shadowy realm of Hades; what is eternal must be of the gods.

The Greeks realized that identity and individuality were such profound concepts, Pesic continues, that they might well be beyond human understanding:

> The mark of identity has a divine indelibility that can only be witnessed, not comprehended. Telemachus knows somehow that he must follow his friend—the deity in disguise—with careful steps. If the god is gracious, Telemachus will find his father and

also himself. The mystery of individuality may be a sacrament that seeks communion.[3]

But this Greek understanding of identity and individuality was just the beginning of the West's reflection on this complex and contentious subject. Pesic's examples include the sixteenth-century story of Martin Guerre, a famous case of imposture; Shakespeare's numerous comedies involving doubles and impersonators; and novels by Dostoyevsky, Conrad, and Kafka—as well as philosophical and scientific arguments advanced by Leibniz, Newton, Davy, Dalton, Maxwell, and others.

Of all features of the quantum realm, the identity of particles—the forms that they can take—is surely among the most bizarre. There are only two possibilities: identical objects that can mash together (bosons) and identical objects that cannot (fermions). Bosons obey a mathematics called Bose-Einstein statistics, while fermions follow Fermi-Dirac statistics. These two possibilities were found independently in 1924–25 with Satyendra Nath Bose and Albert Einstein discovering the properties of bosons, and Wolfgang Pauli's exclusion principle articulating the basic behavior of fermions. As we showed already, and continue to show throughout this book, popular culture often exploits quantum weirdness, finding its terms and imagery packed with creative force. If quantum theory, as Pesic claims, introduces such a radical innovation into our understanding of identity and individuality, why hasn't it revitalized our discussions of these topics? Yet the Pauli principle and Bose-Einstein condensation, representing bizarre forms of identity, are all but absent from popular culture—which is otherwise utterly obsessed with the topic. We haven't yet found it on any T-shirts or coffee mugs. Celebrities do not write about it in their memoirs. No poets we know of invoke it as essential to their work.

THUS, ONCE ALL THE DORM BEDROOMS ARE OCCUPIED BY ROMANTIC PAIRS, ADDITIONAL ROOMMATES ARE FORCED INTO LESS RESTFUL "LIVING ROOM COUCH" ORBITALS.

THE PAULI SEXCLUSION PRINCIPLE

It does not appear as the title of a television show. We have discovered just one cartoon about it, thanks to the ever-inventive xkcd.

Identity is the quantum shoe that hasn't dropped. Why is this most bizarre aspect of a bizarre world not more prominently invoked in a popular culture that celebrates the bizarre and finds meanings for it? To explore why this notion has not yet caught on, we have to look more carefully at the two notions of identity involved.

Wolfgang Pauli and the Exclusion Principle

Wolfgang Pauli was one of the most influential physicists of the 1920s and 1930s, but his influence was often more critical than creative. "Pauli read everything, knew everyone, and wrote thousands of elegantly phrased, passionate, acid-tongued letters to everybody he knew. His criticism was often brutal, a chill wind that occasionally had the effect of freezing out good ideas but killed off many more bad ideas."[4] One of his particularly famous—and devastating—insults was to charge that someone's idea was so bad it was "not even wrong." We regard Pauli, who was full of personal paradoxes, as a symbol for the quantum realm itself: he was simultaneously impish and no-nonsense, quirky and predictable, and indispensable to science but with a flaky side—he corresponded with Carl Jung about his interest in parapsychology and synchronicity.

Pauli was born the same year as the quantum, 1900, to an Austrian family that was a mix of Jews and Catholics. In 1918, two months out of high school, he submitted his first professional physics paper, on an aspect of the theory of relativity; in 1921, two months after receiving his doctorate, he finished a 237-paged definitive presentation of relativity for an important encyclopedia of science that to this day is still considered thorough and insightful.

Pauli was still a young man—all of twenty-five—when he formulated the "exclusion" or "Pauli" principle, for which he would win the Nobel Prize in 1945.[5] A fundamental principle of quantum physics, it is the key to an atom's electron architecture. Because this architecture determines the chemical properties of an element, Pauli's principle is fundamental to chemistry and plays a strong role in the structure of matter. The principle also implies that the electrons and other primal building blocks that it governs are absolutely identical—meaning *indistinguishable*—which is one of the most bizarre aspects of the quantum world.

One of us (Goldhaber) remembers meeting Pauli. In the summer of 1958 I was eighteen, had just finished my freshman year at college in which I had taken a semester of physics, and was accompanying my parents to a physics summer school in Varenna, Italy. Pauli, a participant, was in low spirits. He blamed this on a collaboration with Heisenberg that was going badly. Pauli was now convinced the project was nonsense, distressed that he had taken so long to realize this, and convinced it was a sign he was washed up thanks to his advanced age. "I'm fifty-eight—as old as the century!" he kept muttering. His poor spirits may also have been due to a still undetected pancreatic cancer, which killed him some months later.

The notion that primal bits of matter could be indistinguishable is absent from classical mechanics, which indeed depends on its absence. In classical mechanics, you *have* to be able to put tags on things and follow them around. Each tagged object *has* to have its own past history and its own future course, or Laplace's ideal intellect would not be able to keep track of it. The Newtonian Moment, that is, had a "principle of tag-ability." Sure, atoms and molecules of the same type were thought to be identical—this was essential to make matter stable—but in principle you could track each and every one. Scientists who stopped to ponder the marvelous fact that all atoms of a given kind seem to be identical, such as James Clerk Maxwell, had no answer. There had to be some sort of factory that produced utterly indistinguishable duplicates of atoms, and all he could figure was that it's God's doing somehow.

One hint existed that something was amiss. It came in the form of a puzzle uncovered in the late nineteenth century by the American scientist Josiah Willard Gibbs. Gibbs imagined two adjacent boxes, each containing equal amounts of the same gas at the same temperature. Let's think about what happens if a partition between the two boxes were removed, Gibbs said. The molecules from the two boxes would mix rapidly. If you labeled each molecule by a different number—even numbers, say, for molecules from the first box and odd numbers for molecules from the second—the evens and odds would each start separately but end up mixed together uniformly. This is an irreversible process, and therefore one should expect an increase in entropy, corresponding to decreasing order. But no such increase in entropy occurs, which means that the molecules cannot be distinguished from each other. They don't have separate identities! This startling property makes them different from any macroscopic objects we know—even baseballs, identical twins, and so on. But until the appearance of the quantum (along with the notion of a small number of kinds of "fundamental particles") nobody knew what to make of this.

Pauli's exclusion principle states that no two identical fermions—particles with half-integer spin such as electrons—can occupy the same quantum state. This principle takes for granted the indistinguishability of fermions.[6] "We cannot give a more precise reason for this rule," Pauli wrote in his original article. But what a rule! A fundamental structural principle, it governed all known forms of matter, from atoms to crystals, metals, and chemical interactions.

The *Aufbauprinzip*, he called it: the principle governing how to assemble atoms. Why does this imply indistinguishability? If two electrons were distinguishable—if there were some additional label like color—there would be no reason why they couldn't both be in a state labeled by the same four quantum numbers.

But the exclusion principle governs far more than the structure of atoms, or why electrons occupy the positions they do around atoms. It and the electromagnetic force studied as we saw in the Interlude to Chapter 1 by Charles Coulomb—and the wave properties that prevent each individual electron from being totally localized—govern the shape of matter; why atoms and molecules stay certain distances from each other, and resist being mashed together. Because of these forces, electrons have trouble being pushed into proximity.

Let's imagine, for instance, trying to push two helium atoms into one another. Each has a nucleus indistinguishable from that of the other atom. Each has two electrons (with opposite spin) occupying the lowest possible energy state centered on its nucleus. If the two nuclei come close together, half the electrons have to rise in energy, because they can't all four occupy the lowest state centered on the two coincident nuclei. This requirement, combined with the fact that the two positively charged nuclei repel each other, implies that enormous force is required to push the nuclei together, and even if this were achieved the electrons would still have to be spread out. Thus, the fundamental reason two people cannot be pushed into the same space without mangling them beyond recognition is that the electrons in their atoms can be pushed together only by tremendous, inherently destructive pressure.

The implications of the exclusion principle for identity and individuality are therefore exceedingly bizarre. On the one hand, the exclusion principle implies that electrons have no identities; two electrons in two different states are completely interchangeable. On the other hand, the individuality of the pieces of our world—

any piece of matter as we know it—has been built up thanks to this lack of identity. Individuality in the macroworld is somehow made possible by lack of individuality in the microworld.

Satyendra Nath Bose, Einstein, and Bose-Einstein Statistics

If the exclusion principle is strange mainly because of the associated idea of electron indistinguishability, the possibility found by Bose, that is, any number of indistinguishable particles can be in the same state, and the only label governing this configuration is the number of particles in that state, gives a situation that at first sight has no classical analogue at all. Now any number of particles (not just two) can be in the same "place"! For example, two or fifty or a billion photons could be all within one wavelength of all the others. Those who remember the 1950s will recall a popular college stunt in which a Volkswagen Beetle door opened and an incredible number of students came out, one after the other. Watchers might find it hard to imagine how so many could have piled into that small space, but still would be confident that it was by their contorting themselves rather than having two or more with overlapping bodies in exactly the same volume. Photons may seem to violate that classical intuition. However, by the time you get to a billion or more photons within one wavelength, there is another way to perceive it—as a classical electromagnetic wave. In other words, all these identical particles "merge" with each other so that their particle quality is nearly invisible, and instead they form a wave with an oscillating electric field that can be detected through the force it exerts on electrically charged particles, that is, a classical electromagnetic wave.

The idea came from the work of a Bengali physicist named Satyendra Nath Bose. Bose was born on New Year's Day 1894,

Satyendra Nath Bose (1894–1974)
viewing a photo of Albert Einstein.

not far from what was then known as Calcutta. Bose, who had six sisters and no brothers, was an extraordinary character. Where Pauli had benefited from the most advanced training in theoretical physics that could be found anywhere in the world, Bose made his way in a backwater of the new twentieth-century science, and even of university programs. Even more astonishing, he was only one among several extraordinary students in Calcutta who became known as "The class of 1909," including Bose's lifelong friend and colleague Meghnad Saha. On almost every exam Bose was first and Saha second. In 1914, Bose was married to an eleven-year-old girl, with whom he eventually had nine children.

Once Bose and Saha completed their undergraduate and master's studies in Calcutta there seemed to be nowhere to learn the new ideas emerging from Europe. At the time, India was in the early stages of anticolonial efforts. Bengal was an important center for this movement, because cynical British maneuvers to "divide and rule" had been conspicuous and backfired; these included the decision literally to divide Bengal into eastern (predominantly Muslim) and western (predominantly Hindu) parts. The anticolonial movement was important to Bose and his associates, but they had to be careful to protect their careers, especially since there were few places where they could teach. Somehow, they both were able to find sponsors who literally created new positions for them.

Meanwhile they learned from what foreign books and journals they could find. There were no suitable physics instructors in India even in a relatively advanced region like Bengal. As one way to get around this, Bose and Saha studied and translated into English a collection of articles on relativity, and Bose went on to translate and publish a book by Einstein on general relativity. He and Saha were learning the way music students sometimes do when they transcribe the work of virtuosos.

In 1924, when Bose was thirty, he gave a lecture trying to explain Planck's law for electromagnetic radiation in thermal equilibrium at a given temperature by assuming the radiation consisted of particles (or "quanta" to use the original term), as proposed by Albert Einstein in 1905. In doing this he made a "mistake"; he assumed that a configuration in which many photons occupy exactly the same state is unique, rather than each one of the photons' being labeled by some extra "tag."

He got the Planck answer, and today when physics students are introduced to the Planck law it is through Bose's derivation. As we have seen, that was at least the third derivation, starting with Planck's original argument in 1900, and then Einstein's argument from thermodynamics of photons interacting with molecules in 1916–17. Once Bose realized what he had done, he decided that it was not a mistake at all, but a correct way to count in the quantum world.

Everyone (even Einstein) had taken for granted that counting a bunch of particles in a certain state would include an extra factor for possible rearrangement of the labels of those particles. Bose said there was no extra label; the number of particles in that state was all the information one could have. By this single stroke, although he didn't mention it explicitly, he resolved the Gibbs paradox (as Pauli soon did, also tacitly, for his counting of electron states). While Pauli took for granted the indistinguishability of electrons, Bose took for granted in his counting of states the indistinguishability of photons. Thus both Bose and Pauli assumed an indistinguishability of the particles they considered, just as Maxwell and Gibbs had done for molecules in the nineteenth century for cases where classical mechanics described well the possible configurations of a gas. Indistinguishability can apply for quantum or even classical motions of particles such as small molecules, provided the inner structure of these particles is governed by quantum mechanics.

Bose wrote his paper in English and submitted it to a leading British physics journal, the *Philosophical Magazine*. It's not clear whether the work was rejected or ignored; he received no answer. But Bose felt confident, and daring, enough to send a copy to Einstein himself. Impressed, Einstein translated Bose's paper from English into German and sent it to the leading German physics journal *Zeitschrift für Physik*, together with a note saying this was an important advance, and that he—Einstein—himself planned to submit further work on the subject. Einstein already had reached the height of fame that made him the first and still greatest physics rock star. The journal quickly published the paper—and soon another by Bose that Einstein also translated and submitted, together with a comment that amplified some of Bose's arguments but critiqued others.

Why the Undropped Shoe?

Identity is completely different in the macroworld and microworld. Let's look at the size of leaps in the two domains. The average human can jump up about 2 feet from a standing start, the average flea about a foot. Given the vast differences in body mass involved, it's about the same height. Why? The energy that determines jump height is produced by muscles contracting. If the length that muscle fibers contract is proportional to their length, and the number of muscle fibers and hence the muscle force is proportional to the area across their midsection, then the energy is proportional to the volume of the animal, and hence to the mass of the animal (the density of all animals is pretty much the same). The jump height multiplied by the mass of the animal is proportional to the energy needed for the jump. That means the mass cancels out, and so all animals can jump about the same height. Thus there is a kind of commonality for animals, in that we all are made of the same stuff, but each of us has a different arrangement of that stuff.

All electrons, too, "leap" the same amount in response to the same energy in the same circumstances. But this is because of a completely different reason. An electron has no muscles—in fact no inner structure at all except for its spin. It is not "made" of anything. That is why two electrons can be identical, while two fleas with their many different molecules cannot. That's also why an electron cannot really jump—it has to be kicked!

There is no classical analogue to identity in the quantum world. Both fermions and bosons are mind-bendingly bizarre from the everyday point of view. On the macroscopic level, fermion identity is like what would happen if fraternal but not identical twins were able to inhabit the same space. Boson identity would be like that old vaudeville gag of many people emerging from a small space, like a phone booth or automobile—only an infinite number of people could do this. For both fermions and bosons, quantum identity is a strict application of the saying, "If you've seen one, you've seen 'em all." Particles on the microscopic scale, even up to atoms and modest-sized molecules, fit this description perfectly. If you have seen one spin-up electron, you know that every other spin-up electron will not just *look*, but *be* exactly like the first one, except that the two must be in different places, where "place" means a state with a set of labels.

———

"The radical loss of individuality discloses a new depth behind things," Pesic writes in *Seeing Double*. It is so different that he insists on distinguishing individuality, or what makes an individual be an individual; identity, or the continuing self-sameness of something, which is not the same thing; and what he calls identicality, or the way that members of a species such as electrons "have identity only as instances of that species, without any features that distinguish one individual from another." Things that have identicality have no individuality, no primitive *this*-ness, nothing that would allow saying, "I am me!"

The fact that, as Pesic says, "human individuality rests on anonymous quanta," though "strangely beautiful," may be just too grotesque for human beings to swallow. We love equality, but we hate identity.

Wolfgang Pauli and the Exclusion Principle, Satyendra Bose and Bosons

When Pauli accepted the Nobel Prize at the end of 1946, he wrote of the "shock" he experienced when coming to know of Bohr's quantum theory. There were two approaches for coping with the shock, he said. One was external, starting with the classical language one has already, and treating the quantum world as if one were speaking a foreign language, "looking for a key to translate classical mechanics and electrodynamics into quantum language." This was Bohr's approach. The other, that of Pauli's teacher Sommerfeld, was internal, to look for some cue in numbers that would provide clues from the inside, as if cracking a code.[7]

Pauli hoped to find a way that would enable him to do both at once. The area in which he decided to search was the periodic table. The table arranges chemical elements by properties in groups of 2, 8, 18, 32, as if the electrons came in shells that had the mysterious form $2n^2$, where n has integer values. Why was this? In 1922, Pauli heard Bohr give a lecture in the course of which he asked why all the electrons didn't occupy the lowest shell, known as the K-shell. What was special about the number 2, and then the number 8?

Pauli then considered a view of the German American physicist Alfred Landé in which Landé, who was seeking to explain a phenomenon known as the anomalous Zeeman effect, proposed a rule that an atom with one outermost electron could have only two values of a quantity called R. One physicist called Landé's rule, which came out of nowhere, "completely incomprehensible" even if it allows one to handle "extensive and complicated" phenomena.

The physicist-historian Abraham Pais called it "one of those marvelous moments in science when the lessons of logic are in conflict with the lessons of the laboratory." Pais said, "Landé's formula in fact demonstrates excellently how, in the days of the old quantum theory, gifted physicists were able to make important progress without quite knowing what they were doing."[8]

Pauli wrote a paper on the Zeeman effect in December 1924, in which he said that the two-valuedness was a property not of the "core" but of the outer electron itself. The Zeeman effect, Pauli wrote, arose from "a peculiar not classically describable two-valuedness of the quantum theoretical properties of the valency electron." Pauli's paper of 1925 then generalized that thought, applying it to all electrons. The key sentence: "There never exist two or more equivalent electrons in an atom which, in strong fields, agree in all quantum numbers . . . If there exists in the atom an electron for which these quantum numbers (in the external field) have definite values, this state is 'occupied.'"[9]

The German-born US physicist and philosopher Henry Margenau was one of the few to suggest that quantum identity or indistinguishability should interest more than physicists. In 1944, the year before Pauli won the Nobel Prize, Margenau wrote an article in the journal *Philosophy of Science* which began, "It is strange to note so little discussion of the exclusion principle in the philosophical literature."[10] It must be because of the news pileup, Margenau decided. While relativity was discovered when "little else was going on," the exclusion principle appeared "amid a frenzy of factual discoveries," during the early stages of the revolution that would culminate in quantum mechanics. It was also because the philosophical implications of relativity were obvious—for they implied profound changes in our notions of space and time. If relativity had smashed previous concepts, the impact of the exclusion principle was constructive. The periodic table had been assembled and organized on the basis of chemical properties, and the

Wolfgang Pauli (1900–1958). This picture was taken by Roy Glauber, then a postdoctoral student at Zürich, later to become a Harvard professor and the 2005 Nobel laureate in physics. The circumstances surrounding this picture are amusing. Pauli, Glauber, and other faculty and students were taking an excursion in the hills around Zürich in summer 1950. Glauber was lugging a heavy, bulky camera that used sheets of film, but had only a single sheet left and was reluctant to use it up. "Pauli spent the entire time teasing me about my taking no pictures," Glauber recalls. Eventually the group descended to a flat area on the bank of Lake Lucerne and began to kick around a soccer ball. Corpulent and clumsy, Pauli managed to kick it into the water, and a student had to partly undress and swim to retrieve it. This made Pauli laugh, and provoked him to kick it into the lake again and again, laughing heartily each time. "His roaring laughter became enough of a spectacle that I thought the scene merited my last piece of film," Glauber remembers. "I beckoned for the players to send the ball near Pauli once more." Glauber aimed the camera carefully and pressed the button—just in time, for the soccer ball struck Glauber in the face an instant later.

exclusion principle explained this organization on the basis of the electron architecture.

But the lack of attention paid to the exclusion principle by non-physicists was also partly because of its complexity; you have to grasp some of the new physics to understand. Margenau therefore set out to try for the benefit of his colleagues. You know that matter consists of basic entities called "elementary particles," don't you? These include (to date), he reminded them, electrons, protons, and neutrons, which are stable particles (in matter, at least), but also positrons, mu mesons, and neutrinos, much less visible in our world. All these particles, both familiar and rare, have a common set of properties, one of which is that *they all obey the exclusion principle*, meaning that no two such particles can be found in the same state.

"So what?" he imagines his readers asking. How could they be? Nothing in the Newtonian world can, either—two planets cannot occupy the same space in their orbits, or two billiard balls the same place on a pool table. Duh.

OK, Margenau admits, it's an "incoherent terminology." The quantum world has no such thing as orbits, he wrote for the benefit of those whose atomic education stopped at the Bohr atom. Instead, it has "states." States are strange things. For an electron bound in an atom these states are specified by a set of numbers, called "quantum numbers." Four quantum numbers specify a state, and that's it—nothing else. *Not* four and on top of that the electrons can still be (say) pink or green or blue. Pauli recognized this as just a fact. Only with the advent of full quantum mechanics would this be understood as a deduction: four labels give you a complete enumeration. Furthermore, a state is not even remotely like an orbit but is "a probability distribution for finding the particle at the various points of space." When the exclusion principle says that two electrons cannot occupy the same state, Margenau continues, it means that two electrons cannot have identical sets of quantum numbers.

The idea of a state rather than an orbit cuts down the number of possibilities. To put it somewhat differently than Margenau, think of it this way: on a roulette wheel you have a number of slots where balls can fall onto the wheel. Assuming the wheel is fair and not fixed, a ball is equally likely to fall into any one. Now suppose the balls are identical, so that if you exchanged one with another in a different slot it did not affect the outcome. It would be the "same" state. Pauli's principle was like saying only one ball—well, two, each with opposite spin—could land in such spaces. This vastly shrinks the number of possible configurations of balls in the wheel. For example, you couldn't have five balls in one slot.

What's confusing, Margenau says, is this: physicists are still struggling against "the disappearance of the visualizable traits in their cherished classical theories," and continue, in "sentimental revolt" to indulge the "harmless pleasure" of referring to orbits and perhaps even imagining such orbits. This throws off the philosophers: "How is the philosopher to know what the physicist means when he oscillates between two universes of discourse?"

But the bottom line of the Pauli principle is that "electrons must either have different spin directions or occupy different states." Margenau added that Pauli did not himself use the term *spin*, which was discovered a few years later and put in place of a related concept used by Pauli. Margenau continued with certain caveats. First of all, Pauli's principle applies only to certain kinds of particles, now called fermions. Bosons (which include photons) have no problem crowding into the same state.

Still, Margenau imagined his readers thinking, what's the big deal? The exclusion principle sounds like a residency requirement—a building code for the subatomic world, something imposed or an irrational feature of nature. But it's far-reaching! All at once, the shell structure of atoms becomes clear—not from studying the properties of atoms, but arising from first principles of quantum mechanics. Bohr's spectroscopic observations are explained. So are many of the magnetic properties of solid materials.

Margenau follows the above qualitative explanation with a quantitative, mathematical one. In the Newtonian world, you can state the position and velocity of a piece of matter with six variables. The state is specified by six numbers, and each is a continuous function whose behavior can be described by differential equations. But this is not true in the quantum world. Because of the Heisenberg uncertainty principle, a particle can only be in a "chunk" of phase space. And the exclusion principle says that only one of a set of indistinguishable particles can be in such a chunk at a time.

Why? Margenau finally asks. There's no answer. This is simply a rule, but a very deeply ingrained one. "It is not merely another law of nature, another theorem of quantum physics; it is a regulatory factor directing all reasoning about atomic entities." The ingrained attempt to picture particle behavior in terms of Newtonian orbits makes it *seem* as though there is a lack of causality; the fact that our picture of particle behavior lacks causality results from our defective imaginations, not from nature. What is happening is still deterministic, for states are predictable with exactness. But states are no longer observable, instead determining overall probability distributions of possible events. Observable events "are no longer enmeshed in a deterministic network," Margenau writes. "Like mushrooms, they may crop up anywhere."

Bose's rule about counting photon particles yields the same blackbody radiation formula that Planck had exploited in his earlier derivation; he had found a way to make the particle description of light agree with the wave description of light. Einstein realized the implication: If the number of photons with a given frequency and wavelength is made very large, so that the number of photons in a box one wavelength on a side became large, then it should be possible to describe this system as a classical Maxwellian electromagnetic wave. The photons have "condensed" into something well described as a wave, and the electric and magnetic fields of the wave can be detected from the behavior of electrons interacting with the wave.

Einstein's additional work applied Bose's counting to massive particles, like atoms. Einstein showed that if one could neglect the forces between atoms, then a large collection of such atoms contained in a box would have a lowest-energy configuration in which all atoms were in the same state. An immediate consequence is that at zero temperature all the particles would be in the lowest possible energy state, which therefore is an example of an unproven "theorem" by the German chemist and physicist Walther Nernst: at zero temperature there is zero entropy, that is, no uncertainty whatsoever about the states in which particles are found. This conjecture had long appealed to Einstein, but also puzzled him, and he found it satisfying that Bose's quantum ideas accounted for Nernst's theorem.

Einstein went further: by considering what would happen at finite temperature, he concluded that at a fixed density of the particles there would be a critical temperature below which the gas would condense into a configuration where more and more of the particles are in the lowest energy state. Such a form today is called an Einstein condensate, or a Bose-Einstein condensate. The first example, which took quite a long time to be recognized as an example, is superfluid helium, which exhibits many peculiar properties for macroscopically visible quantities of the helium. If the superfluid is poured into a beaker, it begins to creep up the sides of the beaker, and once at the top glides down to the outside world. No ordinary liquid could do this. The phenomenon—which is quite dramatic to see in person or even on YouTube—is known as a quantum siphon or the spontaneous siphon effect, and is a product of the fact that the entire helium blob is described by a single quantum mechanical wave function—it's a "macroscopic" wave function.

Superconductivity is another example. This one is trickier because the electrons traveling around in a metal are fermions, obeying the Pauli exclusion principle. However, when electrons pair up, they can produce the kind of condensate Einstein discovered, and make "supercurrents" that persist indefinitely even when no voltage is supplied to drive them. Again, it's a macroscopic effect coming from large regions of Einstein condensation.

In 1926, the English theoretical physicist Paul Dirac wove Pauli's remarkable work, and an extension of his ideas by the Italian physicist Enrico Fermi, together with Bose and Einstein's results, couching all in the language of the new quantum mechanics. Dirac's paper sorted types of elementary bits of matter into two: First, particles governed by the exclusion principle, which Dirac called fermions; and second, particles able to "pile into" the same state in arbitrary numbers, or bosons. Crudely, if you exchange two identical objects and the phase factor of the wave function— more on this in a moment—remains the same (it's symmetric), the objects can occupy the same space (there can be a possibility of condensation) and are bosons. If the factor is negative (antisymmetric), the objects cannot occupy the same space and are fermions. Dirac noted that "no other types [of indistinguishable particles] are found in nature."[11] Bosons serve as force-carrying particles— photons, for example, are particles corresponding to electromagnetic forces. As for fermions, Pauli's exclusion principle and the uncertainty principle regulate the atom's energy levels, which determine the chemical properties of elements and the structure of matter. From the viewpoint of our ordinary world, what's even stranger than particles that obey the exclusion principle are those that do not! Both types though, have the quantum-world property of being indistinguishable in principle, something with no classical analogue.

Einstein's prediction of such condensation for massive particles, such as helium atoms, was verified in the laboratory less than fifteen years later. He failed to recognize this as a vindication of his condensation theory, perhaps because he was considering particles that felt no forces among them. In fact, helium atoms repel each other because of the Pauli principle, which doesn't like more than two electrons (one with spin up and one with spin down) to sit on top of each other. Nevertheless, the temperature at which condensation starts is fairly close to what it was in Einstein's calculation, and today this is accepted as the phenomenon he predicted.

As with light, a condensate of helium can show wave coherence over large distances, but there is a difference: The absolute phase of the wave cannot be observed, as can the direction of the electric field of a laser light wave. In other words, there is no completely classical wave limit of the Einstein condensate.

In the end, Margenau's speculation that the reason the public shows little interest in particle identity as an aspect of quantum mechanics might be due to the fact that "so much else was going on" at the time seems unpersuasive to us. Sixty years later, and even with explanations like his and Peter Pesic's, quantum identity still does not resonate outside the specialized fields that utilize quantum mechanics. We wonder if the explanation is rather that the notion is simply unpalatable; people (nonquantum mechanics, anyway) don't like the radical conformity implied by the notion that there is no difference whatsoever between two entities of the same type.

Chapter Six

Sharks and Tigers: Schizophrenia

Quantum Man is a sculpture by Julian Voss-Andreae installed in the City of Moses Lake, Washington. Vastly different from *Quantum Cloud*, the steel-wool-like sculpture by Antony Gormley discussed in Chapter 4, this one is made of parallel steel sheets 2.5 meters high, and it changes in form as you walk around it. From one perspective it reveals the outline of a human being, while from another the human form disappears entirely. Voss-Andreae calls his sculpture a metaphor for wave-particle duality, or the way in which a quantum phenomenon can appear as either a particle or wave depending on how we look at it.[1]

Quantum Poetics, a book by Harvard English professor Daniel Albright, seeks to understand Modernist poetry—the term *Modernist* reserved for those poets who wrote during the half century or so before the Second World War and who tried to make a clean break with their predecessors—by mapping its authors onto wave and particle models even when their poetry does not mention physics. Ezra Pound was "the Democritus, the Rutherford, the Bohr of poets," a researcher into the elementary particles of poetry, while D. H. Lawrence was a proponent of "feeling-waves, the radiations that interconnect man, woman, snake, cow, moon, sun, through the complicated electromagnetic attraction and repulsion of solar plexus and lumbar ganglion." Yet, Albright continues, both of these poets—and all Modernists—recognized that the behavior of poetry does not fall neatly into either the particle or wave model. He concludes that Modernist poets sought to maintain both con-

Quantum Man by Julian Voss-Andreae.

tradictory models: "the Modernist poets teach themselves how to conceive the poem according to the wave model and the particle model *at the same time.*"[2]

Voss-Andreae's sculpture and Albright's Modernist poets are creative because they are two things at once. Both people see this productive schizophrenia as embodying a lesson of quantum physics. But sometimes one simply jokes about quantum schizophrenia. We once saw a cartoon, for instance, picturing the offices of the "Universe Corp R&D Division," in which a bright young engineer, dressed in an immaculate white shirt and tie and holding a coffee cup, approaches God with his latest innovative idea: ". . . and if you could make it a wave AND a particle, that would be great!" God is listening with His usual caring patience, but His thought balloon shows Him thinking, "Christ, what an idiot!"

The Newtonian world was not at all schizophrenic. Everything in it belonged to one of two "bins." In one bin were particles/corpuscles, whose masses were found at specific places, were pushed and pulled by forces, and always had a definite momentum and position. In the other bin were waves, which were described by Maxwellian theories that used continuous functions, obeying differential equations, to depict processes that smoothly evolve in space and time. Both theories involved observable and predictable phenomena. You input information about the initial state, run the program, and out pops a prediction of a future behavior.

Quanta belonged in which bin?

Not clear. During the early to mid-1920s, in the latter days of the first quantum scientific revolution, physicists tended to be partisans of one bin or the other, trying to stretch either the theory of particles or of waves so that one size fit both. They were trying to figure out what kind of animal, so to speak, quantum phenomena were, for what they saw of these phenomena made them appear like quite different animals, which had to be wrong. Then, in the second quantum scientific revolution of 1925–27, physicists figured out how to put waves and particles together. The outcome, stranger than schizophrenia, was that quantum phenomena *were* indeed all these different kinds of animals, depending on your approach.

Artists, writers, and various kinds of scholars adopted this outcome as revealing the reality of things with necessary but incompatible aspects, and as an indication that it was legitimate to have contradictory perspectives about an issue. Religious writers seized on the idea as a scientific analogue of the way Christian believers think. The physicist John Polkinghorne, for instance, wrote that when Bohr said anyone who claimed fully to understand quantum physics thereby reveals that they have not yet begun to comprehend it, he was echoing a remark by the Church of England Bishop William Temple, who remarked that "if any man says he understands the relation of Deity to humanity in Christ, he only makes it clear that he does not at all understand what is meant by Incarnation."[3]

Polite but Determined Lunatics

We have to appreciate how contorted the first quantum revolution left things in order to grasp the magnificence of the second.

In the early 1920s, relativity and quantum mechanics were often mentioned in the same breath as two great triumphs of science. But many scientists also had difficulty comprehending the breakthroughs. In 1921, in an article entitled "Oh, Quanta!" the renowned American physicist Arthur G. Webster of Clark University—founder of the American Physical Society, the US organization for professional physicists—wrote about Planck's having received the Nobel Prize the previous year (Planck won it for 1918, but the celebrations were delayed due to World War I). Webster, who had devoted his career to classical physics, summarized the history of the quantum as best he could, then added:

> To understand the theory of quanta requires a knowledge of all the most difficult parts of mathematical physics. I do not half understand it. Do you? But, like the theory of relativity, it is a great thought, worthy of the Nobel Prize.[4]

Webster committed suicide shortly thereafter, amid personal problems that included possible closure of his department and despair over his physics contributions.

If mainstream physicists like Webster had trouble digesting the quantum, the public had more. A few nonphysicists made zany, tongue-in-cheek, or uninformed speculations about the meaning of the quantum. An Indian spiritualist and astronomer named G. E. Sutcliffe told the *Times of India* that relativity and quantum mechanics heralded the downfall of Western science and the supe-riority of Eastern doctrines. "The reply of the west to the sphinx-n riddle destroys the laws of Newton, the eastern reply keeps them intact," he wrote, cryptically.[5] A year later, the *Los Angeles Times* joked that if this new thing the quantum is indeed a powerful new form of energy, it might finally provide enough force to hold the Democratic Party together.[6] (Today, the *Times* surely would have mentioned the Republicans instead.)

Scholars in other fields who had confidently relied on Newtonian analogies now reassessed. The philosopher and psychologist F.C.S. Schiller found that atoms, whose transitions now had to be triggered by certain minimal transition energies, behaved like certain psychological reactions—meaning that "Psychology may have light to shed even upon Physics."[7] A few years later, the Harvard historian William Munro told colleagues that Walter Bagehot's attempt to ground democracy in Newtonian mechanics, with its postulate of "a series of ultimate and fixed uniformities," needed revising.[8] But these were isolated and random speculations, scholarly shots in the dark. The general reaction of scholars and everyone else was that quantum mechanics was like relativity—something only a handful of scientists understood.

In 1922, Charles Darwin—the physicist grandson of the famous naturalist—spoke about quantum theory to the Ebell Club, a Los Angeles women's group that hosted prominent guests and whose events were publicized. Darwin, from Cambridge University in England, was about to spend a year at the California Institute of Technology in Pasadena. The day after his talk, a *Los Angeles Times* article opened with words from an overheard conversation:

"I don't understand a word he is saying."
"Never mind, you can read all about it in The Times tomorrow."

Myra Nye, the reporter, confessed that she couldn't help her readers and that none of the lunch guests had understood the theory. "When Charles Darwin, long ago, first advanced his evolution theory he could have created no greater bewilderment than did his grandson yesterday." But Nye realized that quantum theory was truly bizarre. Darwin, she said, had envisioned an eminent scientist of 1895 falling into a trance, awakening today, and learning of recent physics. That Rip Van Winkle would be "depressed" by the strangeness of the theory of relativity, Nye reported Darwin as saying, "but being a rational and intelligent man he would accept it." However, if someone tried to tell him about quantum theory, "he would think quite definitely that his informant was raving."[9]

A few years later, the *New York Times* ace science reporter Waldemar Kaempffert covered a talk given by Heisenberg at a meeting of the British Association for the Advancement of Science. Heisenberg was not yet a celebrity; this was his first public talk on his soon-to-be-famous, hard-to-explain principle. Heisenberg's theory, Kaempffert wrote, "will make it necessary to modify belief in what we are pleased to call 'common sense' and 'reality.'" But the reporter could not figure out what replacement Heisenberg was offering:

> The layman without a knowledge of higher mathematics, listening to Dr. Heisenberg and those who discussed his conclusions, would have decided that this particular section of the British Association is composed of quiet and polite but determined lunatics, who have created a wholly illusory mathematical world of their own. The conception is that they and their kind alone have a proper view of "reality"; the rest of us live in a dream world fashioned by ill-understood words.
>
> To explain the quantum theory and its modification by Dr. Heisenberg and others is even more difficult than explaining relativity. It is much like trying to tell an Eskimo what the French language is like without talking French. In other words, the theory cannot be expressed pictorially and mere words mean nothing. One is dealing with something that can be expressed only mathematically. . . .
>
> [T]he whole science of mechanics must be rewritten. And when it is rewritten, no one but a mathematician will be able to understand it.[10]

The difficulties with understanding quantum theory, however, went beyond complex math and included elements that puzzled scientists themselves. Physicists, it would turn out, were starting off on the wrong foot. They were starting with a classical approach to things like action and phase space, then trying to find the right rules to adapt it to the quantum realm. At the end of its first quarter-century, it was a conceptual mess. Jammer called it "a lamentable hodgepodge of hypotheses, principles, theorems, and computational recipes rather than a logical consistent theory."[11] Many scientists still hoped the quantum would go away.

The quantum's most disturbing implication was that the sub-atomic world was thoroughly schizophrenic. Quanta were wave-like, and spread out spherically from their source without specific position or direction, broadening and thinning in space and time. But they were also like particles—then called *corpuscles*—each with its own definite position and momentum, always following a specific path in space and time. Neither waves nor particles were baffling by themselves; the fact that quantum things behaved like both was. Newton had noticed wavy properties of light despite having declared that it is made of particles, but never dreamed that these belonged in an integral package. In the quantum world, they were. As Planck put it in his Nobel lecture of 1920, what happens to the light quantum after emission is a puzzle: does it spread out like a wave, or fly like a projectile? If the first, how can it do things like knock electrons out of atoms? If the second, how can it behave electromagnetically according to Maxwell's equations?

In a public lecture delivered in 1921, the physicist Sir William Bragg captured the schizophrenia in a remarkable image:

I drop a log of wood into the sea from a height, let us say, of 100 feet. A wave radiates away from where it falls. Here is the corpuscular radiation producing a wave. The wave spreads, its energy is more and more widely distributed, the ripples get less and less in height. At a short distance away, a few hundred yards perhaps, the effect will apparently have disappeared. If the water were perfectly free from viscosity and there were no other causes to fritter away the energy of the waves, they would travel, let us say, 1,000 miles. By which time the height of the ripples would be, as we can readily imagine, extremely small. Then at some one point on its circumference the ripple encounters a wooden ship. It may have encountered thousands before that and nothing has happened, but in this one particular case the unexpected happens. One of the ship's timbers suddenly flies up in the air exactly 100 feet, that is to say, if it got clear away from the ship without having to crash through parts of the rigging or something else of the structure. The problem is, where did the energy come from that shot this plank into the air, and why was its velocity so exactly related to that of the plank which was dropped into the water 1,000 miles away?[12]

Bragg studied X-ray diffraction in solid crystals, which scattered a beam of X rays coming from a source (the dropped "log"); these rays were diffracted by the structure of the target crystal (like the wave), and then detected when they produced a dark spot on an emulsion (the "timber").

Einstein was a partisan of the particle bin. In 1916, as we saw, he extended his idea that light consists of physically real quanta, each having a particular direction and momentum. This process conserved energy, for the amount emitted equaled the amount absorbed. Overstating things somewhat, he had proclaimed that "radiation in the form of spherical waves does not exist." But he had to incorporate statistics in his theory to make it work in the form of "probability coefficients" that described the emission and absorption of quanta. He found this painful but hoped it would soon be replaced by a deeper understanding. Einstein's allies included Arthur H. Compton, who in 1923 discovered the "Compton effect," that when photons bounce off electrons they come from, and rebound in, definite directions. Definitely particle behavior![13]

Darwin, the physicist who had addressed the Ebell Club, was a wave partisan. He was disturbed to discover the price he would have to pay to revamp wave theory to cover quanta. Wild ideas are needed, he wrote, and sarcastically suggested that physicists might have to "endow electrons with free will."[14] The least wild thing, he finally decided, would be to keep wave theory by abandoning conservation of energy—which ever since the middle of the previous century had been a bedrock of physics—for individual events.[15]

Many other scientists tried their hand at finding a satisfactory way to convey the necessity yet also irreconcilability of wave and particle theory in ordinary language. British physicist C. D. Ellis compared waves and particles to the plan and elevation of an engineering drawing, which "reflect our attempt to describe in two dimensions a three-dimensional object."[16] J. J. Davisson spoke of differences between the behaviors of rabbits and cats, J. J. Thomson of a struggle "between a tiger and a shark, each is supreme in his own element but helpless in that of the other."[17] H. S. Allen described two disconnected buildings, with physicists forced to live now in one, now in the next.[18] The situation resisted metaphor because the everyday world had nothing like it. Einstein wrote despairingly, "There are therefore now two theories of light, both indispensable and—as one must admit today despite twenty years of tremendous effort on the part of theoretical physicists—without any logical connection."[19]

The quantum was ruining the work of the best and brightest. James Jeans ended a series of popular lectures by saying, "I know of no phenomenon in the whole of physics which helps us in the least to comprehend the physical processes at work. With this fact before us, not much meditation is needed to convince us that we are still very far from understanding the working of the atom or the true meaning of atomicity and quanta."[20]

In his 1921 public lecture Bragg was blunter:

No known theory can be distorted so as to provide even approximate explanation. There must be some fact of which we are entirely ignorant and whose discovery may revolutionise our views of the relation between waves and ether and matter. For the present we have to work both theories. On Mondays, Wednesdays, and Fridays we use the wave theory; on Tuesdays, Thursdays, and Saturdays we think in streams of flying energy, quanta or corpuscles. That is, after all, a very proper attitude to take. We can not state the whole truth since we have only partial statements, each covering a portion of the field. When we want to work in any one portion of the field or other, we must take out the right map. Some day we shall piece all the maps together.

Heisenberg, Schrödinger, and the Second Quantum Scientific Revolution

Then, in 1925–27, the whole map—in fact, *two* whole maps—were found, in one of the most remarkable and far-reaching developments in the history of science. The two mappers—Werner Heisenberg and Erwin Schrödinger—had not tried to adapt the classical map but had started over from scratch. The trouble was, the two maps were completely different and the territory they mapped—the true nature of atomicity and quanta—was more bizarre than anyone anticipated.

If the first quantum revolution involved the astonishing discovery of an alien creature appearing in Newtonian territory—with more and more instances of this alien creature cropping up—the second quantum revolution of 1925–27 involved the even more astonishing discovery: we have been living in alien territory all along! The story of this second quantum revolution, the finding of this new world, itself has become a legendary, oft-told tale.

In 1925, Heisenberg abandoned any attempt to picture the subatomic world. He decided to try to describe the quantum solely in terms of the mathematics of its observed properties, in a "map" called matrix mechanics. This map included a specific limit on what could be known about the position and momentum of observed particles—the uncertainty principle. The next year, the Austrian physicist Erwin Schrödinger described the quantum as a peculiar kind of wave that evolved in time continuously and predictably according to familiar differential equations. But what this wave gave you, it turned out, was not space coordinates but information about the *probabilities* of particles appearing. Their achievements so transformed quantum theory that it brought about a second quantum scientific revolution, called "quantum mechanics" to distinguish it from earlier quantum theory.

After this revolution, quantum mechanics had two parts. One part was a special kind of field that doesn't tell you where a particle will be next, only some information, in the form of probabilities, about where it *may* appear next. The information this field gives you about the world is not unique; there is more than one possibility, and *all* the theory gives you is lists of possibilities. It cannot tell you what the "real" is going to look like, only information about its

"ACTUALLY I STARTED OUT IN QUANTUM MECHANICS, BUT SOMEWHERE ALONG THE WAY I TOOK A WRONG TURN."

possible forms. The second feature, inevitable in any theory based on probabilities, is that something distinct from that field intervenes to bring one of these possibilities into existence.

Werner Heisenberg, born in 1901, was a young man when he did the work that brought him fame and the Nobel Prize. His father was a professor of Greek at the University of Munich. "Heisenberg had the character one often associates with poets: dashing good looks, a physical frailty including severe vulnerability to allergies, excellent musicianship, and a sensitive and often emotional responsiveness to the world around him."[21] He was also wildly ambitious. After finishing his doctorate in Munich in 1923, he embarked on the quest to make the new quantum physics rational.

Heisenberg sensed that the quantum was ill-suited to the Newtonian assumption that events unfold on a four-dimensional space-time stage. On that stage, everything is always at a specific place at a specific time. When things move, they are pushed or pulled by forces of specific strength, and trace out definite paths in space-time. This ontology, to use philosophical language, is visualizable. The attempt to visualize quantum phenomena in this way had already defeated the best physicists of the time, including Planck, Bohr, Einstein, Sommerfeld (Heisenberg's adviser at Munich, 1923), and Max Born (his postdoctoral supervisor at Göttingen). Heisenberg thought the attempt to construct a visualizable solution might be the source of the trouble. Junk the positions and paths and orbits! Visualizability—*Anschaulichkeit* was his German word for it—had to go! So in the first few months of 1925 Heisenberg stopped trying to produce theories that pictured atomic events. He would try to describe frequencies and amplitudes, momentum (p) and position (q) not as classical properties with specific values, but in purely mathematical terms as "operators" on functions.

In May 1925, Heisenberg suffered from such a severe attack of hay fever that it brought his work to a halt. He headed for Helgoland, a rocky North Sea island remote from allergen producers.

Gradually recovering on this island refuge, he returned to work. Sensing a breakthrough, one night he worked late, finally piecing together his ideas by about 3 a.m.:

> At first I was deeply alarmed. I had the feeling that, through the surface of atomic phenomena, I was looking at a strangely beautiful interior, and felt almost giddy at the thought that I now had to probe this wealth of mathematical structures nature had so generously spread out before me. I was far too excited to sleep, and so, as a new day dawned, I made for the southern tip of the island, where I had been longing to climb a rock jutting out into the sea. I now did so without too much trouble, and waited for the sun to rise.[22]

Returning to Göttingen in June, Heisenberg dashed off a paper. Our experiments have implied that we cannot "associate an electron with a point in space," he wrote, suggesting the need "to discard all hope of observing hitherto unobservable quantities, such as the position and period of the electron." This wasn't quite true, but it captured his spirit. He showed how to compile tables with rows and columns of amplitudes and frequencies associated with transitions between states, which he called "quantum-theoretical quantities." He related these quantities by a new kind of calculus: "quantum-mechanical relations." Multiplying the tables with the new calculus, he produced formulas similar to those of classical mechanics—a strange mathematical trick that vastly increased the ability to calculate and predict quantum behavior. The only snag involved something called commutation. The commutative law is the familiar mathematical principle that the order in which one multiplies two numbers does not matter: $ab = ba$. But in Heisenberg's funny calculus this was not true. When he multiplied one quantum-theoretic table (we'll call it A) by another (B), the order mattered: AB is different from BA. After Heisenberg gave Born a draft of the paper, Born saw that Heisenberg had reinvented the wheel; the strange tables were what mathematicians call matrices, and the bizarre calculus was simply the way mathematicians multiplied matrices. Matrices were noncommutative. It was a remarkable case of a purely mathematical development turning out to describe the real world of particle motion.

Born realized that embedded in Heisenberg's work was a marvelous discovery: the specific amount by which the tables of quantum theoretical quantities did not commute. When the matrix **p** associated with momentum and the matrix **q** associated with position (physicists often write matrices in bold to distinguish them from ordinary variables) were multiplied in the two different orders, the difference between them was proportional to Planck's constant. More specifically, Born and his other assistant Pascual Jordan found that $\mathbf{pq} - \mathbf{qp} = \mathbf{I}h/2\pi i$ (**I** is the unit matrix, which may be left out as a factor with no change in meaning, and i is the "imaginary unit," a number whose square is -1). Born would write later, "I shall never forget the thrill I experienced when I succeeded in condensing Heisenberg's ideas on quantum conditions in the mysterious equation $\mathbf{pq} - \mathbf{qp} = \mathbf{I}h/2\pi i$, which is the center of the new mechanics and was later found to imply the uncertainty relations."[23] In February 1926, Born, Heisenberg, and Jordan published a landmark paper in physics that explored the implications of this equation, the first complete map of the quantum domain.

Neither Heisenberg, nor his early readers, realized the magnitude of what was happening. "The mathematics was unfamiliar, the physics opaque," wrote historian Pais.[24] Many skeptics thought, Let's wait and see. Hesitations were dispelled when Pauli applied Heisenberg's rules to the hydrogen atom, deriving the Balmer formula. Still, Pais called it a "highly mathematical technology."

Soon physicists were shocked by a *second* complete map of the quantum domain. The Austrian physicist Erwin Schrödinger had produced a much more readable map based not on the complex mathematics of matrices but on wave mechanics. The mathematics of waves was easier to use, and waves were visualizable. Their properties—frequency, amplitude, wavelength—flowed smoothly and continuously, and the approach seemed to restore the space-time stage and visualizability of subatomic events. Schrödinger's wave mechanics had a puzzling feature called the ψ-function, the stuff that "waved." Born soon interpreted this as saying nothing about the event itself but rather its probability (for which he won the Nobel Prize in 1954). Amazingly, the two maps turned out to be identical. Both used tools of classical mechanics in different ways, including ideas introduced for the kinetic theory of gases among other things.[25]

Heisenberg insisted that the quantum realm was unvisualizable. In an article of September 1926, he wrote that our "ordinary intuition" does not work in the subatomic realm, which has a different kind of reality. "Because the electron and the atom possess not any degree of physical reality as the objects of our daily experience, investigation of the type of physical reality which is proper to electrons and atoms is precisely the subject of quantum mechanics."[26] Quantum mechanics, evidently, was going to upend our notion of reality itself.

A furious discussion ensued between Heisenberg and Pauli that is fascinating to read almost a century later.[27] "But now comes the obscure point," Pauli wrote in October. "The p's must be assumed to be controlled, the q's uncontrolled. That means one can always calculate only the probabilities of particular changes of p's with fixed initial values, [and then only when they are] averaged over all possible values of q." Pauli then covered a page and a half demonstrating that the more you learned about one, the less you could say about the other. "So much for the mathematics," he continued, still perplexed. "The physics of this is unclear to me from top to bottom. The first question is why can only the p's (and not simultaneously both the p's and the q's) be described with any degree of precision?" Annoyed, Pauli summed it up as follows: "You can look at the world with p-eyes or with q-eyes, but open both eyes together and you go wrong." What could this mean?

Discussions, letters, and papers repeatedly spilled from mathematics into physics and philosophy and back again. Perhaps, Heisenberg wrote back, properties like space and time are only statistical concepts, like the temperature and pressure of a gas. "It's my opinion that spatial and temporal concepts are meaningless when speaking of a *single* particle." He continued, "What the words 'wave' or 'corpuscle' mean, one does not know any more."

At this moment, for the first time, the utterly radical nature of what physicists had accomplished became clear. Suddenly, they realized the theory had philosophical implications, for it was clear that their assumptions about reality—the basic features of matter, which are usually left unexamined—were called into question and seemed to require critical reexamination.[28] Now, the quantum seemed to have something to say not only to philosophers, but also to interested members of the public. The public was about to be acquainted with the quantum, with waves and particles, with randomness and uncertainty, and even with equations like $E = h\nu$ and $\mathbf{pq} - \mathbf{qp} = h/2\pi i$. Like it or not, the public would be told, these things are the stuff of which the modern world is fashioned. The Quantum Moment was about to be born.

The Irrational Quantity

On Valentine's Day 1935, the British dramatist and writer John Beresford—great-grandfather of the American actor James Newman—published "The Irrational Quantity" (its subtitle was "$\mathbf{pq} - \mathbf{qp} = \sqrt{-1}\ \hbar/2\pi$"), a fanciful short piece (with a mangled equation in its title), in the *Manchester Guardian*.[29] In it, Beresford describes a dream he had about the weirdness of quantum theory. He describes himself sitting in his bath when what should materialize out of the fragrant steam produced by lavender bath salts but the Square Root of Minus One, "a lop-sided, angular little fellow, blind of one eye, with a habit of keeping an arm crooked behind his head, and he had a little thin voice that slurred suddenly up and down the scale with a metallic buzz and a dying fall, or rise, that left me with a sense of his having covered the whole gamut in a single note. Unquestionably a very queer little fellow." When Beresford asks the creature where he came from, the reply is, "From the mind of man." The creature came to lament that he was being blamed for introducing uncertainty and irrationality into science and the modern world via the uncertainty principle and other aspects of quantum physics. Newton and others kept their mysticism separate from their mathematics, says the Square Root of Minus One. Now, "There's a sort of a hole in modern mathematics and all sorts of queer, irrational things keep dropping through it." The Square Root of Minus One then suddenly announces that he is having a romance with—dropping his voice—"h." The *sotto voce* had the effect of drawing Beresford's attention to the fact that it was the "h" of the uncertainty principle and not, say, Helen of Troy. "I have noticed that you're always about together in a manner of speaking," Beresford says. Indeed, the creature replies, and he was thrilled to say that there is nothing uncertain about her reciprocal love for him: "They call her Planck's constant, you know."

Beresford's vivid image of a hole, created by h and the square root of -1, letting in "queer, irrational things" embodies the realization that quantum mechanics is not just a complex theory but means that the world is populated by distinctively non-Newtonian kinds of objects. If we had some sort of crank that would make h bigger or smaller, and we were able to turn h to zero, the hole would close, and the world would be classical and populated only by Newtonian objects. But we cannot shut the hole, which means not only that weird, irrational things are here to stay, but also that the world itself is full of what Updike called gaps, inconsistencies, warps, and bubbles.

"Obvious fact," writes David Foster Wallace: "Never before have there been so many gaping chasms between what the world seems to be and what science tells us it is. 'Us' meaning laymen. It's like a million Copernican revolutions all happening at the same time." He cited examples including the curvature of space, and quantum particles that are both there and not. "We 'know' a near-infinity of truths that contradict our immediate commonsense experience of the world."[30]

Wallace is wrong, of course. In its domain and context, science describes what happens in the world nicely—that's its job! But even outside that domain and context, via metaphorical mechanisms that we discussed in Chapter 3, it can help us to express our experience of being human. Quantum language and imagery is especially metaphorically appealing because it reflects difficulties we face in describing our own experiences; quantum mechanics is strange and so are we.

In another essay, Updike wrote:

> The non-scientist's relation to modern science is basically craven: we look to its discoveries and technology to save us from disease, to give us a faster ride and a softer life, and at the same time we shrink from what it has to tell us of our perilous and insignificant place in the cosmos. Not that threats to our safety and significance were absent from the pre-scientific world, or that arguments against a God-bestowed human grandeur were lacking before Darwin. But our century's revelations of unthinkable largeness and unimaginable smallness, of abysmal stretches of geological time when we were nothing, of supernumerary galaxies and indeterminate subatomic behavior, of a kind of mad mathematical violence at the heart of matter have scorched us deeper than we know. Giacometti's wire-thin, eroded figures body forth the new humanism, and Beckett's minimal monologuists provide its feeble, hopeless voice.[31]

If only all human voices were as articulate as Beckett and Giacometti's! Too frequently, the use of quantum language and concepts in popular culture amounts to what the physicist John Polkinghorne calls "quantum hype," or the invocation of quantum mechanics as "sufficient license for lazy indulgence in playing with paradox in other disciplines."[32] This is how it principally appears in things like TV programs, cartoons, T-shirts, and coffee mugs. Updike suggests, however, that although we keep asking for the material things that the quantum is going to give us technologically, it has already changed us spiritually. It has provided a range of novel and helpful images to interpret our experiences, and thereby give us a better grip on the world, on a scale comparable to Newtonian mechanics. By heightening our sensitivity to the gaps, inconsistencies, warps, and bubbles of experience, quantum language and imagery have made the world itself different. Literature written in this changed world may reflect the influence of quantum language and imagery even when it does not specifically use it.

Consider, for instance, *Open City*, a novel by Teju Cole, whose protagonist, Julius, is a German-Nigerian psychiatrist who lives in New York City. Julius is prone to taking random walks around his adopted environment, in the course of which his thoughts flit discontinuously from one subject to another: what the city must look like to a flock of geese, the history of the slave trade in New York City, the causes and effects of 9/11, various works of music and literature. His voice is confident, knowledgeable, emotionless, and detached, even as it jumps about in a kind of quantized mental phase space, without any ruminations or meditations knitting the jumps into a smooth fabric, or discomfort that it lacks such a smooth fabric. "We experience life as a continuity," he thinks, "and only after it falls away, after it becomes the past, do we see its discontinuities. The past, if there is such a thing, is mostly empty space, great expanses of nothing, in which significant persons and events float."[33]

While this is as close as Cole's protagonist comes to using quantum language and imagery, it expresses a sensitivity to the warps and bubbles that we notice when we tend to the world carefully, which our Newtonian intuition ordinarily prompts us to gloss over, fill in, or assume to be inconsequential. Many scholars have argued that fractioning and discontinuity, gaps and inconsistencies are a distinctly contemporary part of modern life. In *Turing's Cathedral: The Origins of the Digital Universe*, for instance, George Dyson argues that digitalization reflects a lack of continuity that is endemic to the way we live in the modern world. Many modern novels also reflect this, though in different ways. Several of what might be called "centrifugal" novels, where the characters are refracted around a central focus—Colum McCann's *Let the Great World Spin*, for instance—are examples; these gradually build up a picture, pointillistically. Cole's book—which won the International Literature Prize for 2013 in Germany and the Hemingway Foundation/PEN Award for first novel, and was a finalist for the National Book Critics Circle Award—is of a different sort, what we might call the literature of warps and bubbles, where the gaps and inconsistencies are integral to the whole story; other examples include works by W. G. Sebald and Julian Barnes.

The inconsistencies and gaps of experience were there all along. Quantum imagery and terminology give us a language to describe something we always already knew. Here we have art reflecting its time in a particular way. Not that the quantum either caused this literature nor was a product of the same cultural forces that produced it; still, it happened at the right time to provide vivid imagery for capturing our modern experience.

It is doubtful that Cole or the other authors mentioned thought explicitly about quantum theory while writing their novels; rather, its physics concepts have become so much of an intellectual tool for understanding and describing the contemporary world that people who use it are not even necessarily aware that they are doing so.

It's simply part of the toolkit of any thoughtful writer, and helps to achieve a better grip on experience, and a clearer voice. But as any carpenter knows, the acquisition of new tools brings not only more precision to one's existing work but also the ability to make new things.

In the next chapters we provide examples using specific kinds of language and images.

Erwin Schrödinger's Map/ Werner Heisenberg's Map

In class we go into some detail about the contrasting approaches of Schrödinger and Heisenberg, providing key vocabulary that recurs in discussions. This we do in a way that is accessible even to nonscience majors if they exert a little patience, and we now condense it in this interlude. If you like to snooze in class you can skip this part.

We use two assumptions about light. The first is one Einstein made in his photoelectric effect paper that the energy of a light quantum, or photon, is localized in a small region as if it were a particle, rather than spread out like a wave—although its energy is somehow connected to waves by Planck's formula $E = h\nu$, where E is the energy, h is Planck's constant, and ν a wave frequency. The second assumption is that the *intensity* of these photons, that is, the number per unit time crossing a unit surface area, is related to the light's intensity. The energy per unit time crossing a unit area is called the *power intensity*. Maxwell's theory of electromagnetism gives the power intensity of a classical electromagnetic wave by what is known as a "Poynting vector," an expression that involves both the electric and magnetic field strengths. For a fixed frequency, you get the photon intensity by dividing the power intensity by the energy per photon, or $h\nu$.

Now imagine that a light wave is diffracted, so that very little power arrives at one part of a screen and a lot at other points. Photons, that is, are hitting the latter areas more often than the former. We can also write an expression for the instantaneous energy *density* in a wave, again depending on the electric and magnetic fields. The energy density is simply the energy per unit volume, and therefore the photon number density is the energy density divided by the energy per photon, $h\nu$. Where this density is large, one should detect many photons, but where it is small, there are far fewer. We can even imagine the case of a *single* photon, perhaps trapped in a cavity such as a microwave oven. The total energy stored in the electromagnetic field should be exactly $h\nu$, but the photon is also assumed to be localized. What is going on in this funny case?

In quantum mechanics a *wave function* is written for this single photon, whether free or trapped in a cavity. The *probability density* for finding the photon at some location is proportional to the square of the magnitude of this wave function. That is, if we made repeated attempts to find this photon, we'd find it sometimes in one part of the cavity, sometimes in another. The probability density would tell us how often it would turn up, where.

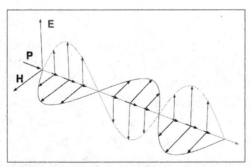

Picture of a classical electromagnetic wave, with E indicating the direction of the electric field, H indicating the direction of the magnetic field, and P (pointing along the direction of wave motion) indicating the power intensity, or energy per unit time per unit area carried by the wave. The power intensity of the wave divided by $h\nu$ gives the number of photons per unit time per unit area. In quantum mechanical applications, when the number of photons in a box the size of a wavelength is small, we can deduce from the energy density of the wave the probability per unit volume of finding a photon near some location.

Picturing what is going on in this way is called a *physical interpretation*. This one illustrates two things: First, the electromagnetic field is completely calculable, its evolution in time exactly predictable. But second, the position of the photon at a given time is *not* predictable, for we know only a probability distribution for its possible appearance. We may call the electromagnetic field the *wave function* of the photon, whose square gives the probability density for finding the photon in the vicinity of a given location. But there is an important difference between the quantum mechanical description of a photon and of an electron. Classical physics gives a law of motion for the photon's wave, via Maxwell's equations, that requires the photon to move at the speed of light. But classical physics does not provide the law of motion for an electron wave; it does not even include the *notion* of an electron wave. Electron particles do not have to move at the speed of light and can be slow-moving or even at rest. Just as Einstein invented the notion of light particles, Schrödinger had to discover the wave function that describes probability density for electron particles.

Schrödinger's Approach

In the winter of 1925–26, Schrödinger set out to develop the analogue of an electromagnetic wave function for particles such as electrons (at least, for slow-moving, nonrelativistic ones) on analogy with the classical electromagnetic wave function for photons. Schrödinger's wave equation—which he famously worked out over the course of a few weeks while holed up with a mistress in the Alps—is completely deterministic, like its classical electromagnetic counterpart. He constructed it out of several key ingredients. One was Planck's relation between energy and frequency, $E = h\nu$. Another was a relation between momentum and wavelength discussed by Einstein but first applied to electrons in atomic physics by Louis de Broglie, $p = h/\lambda$. A third key element was differential equations, as in Maxwell's equations used to relate rates of change of the function in time and position. Because these rates are governed by the frequency and wavelength, this gives rules determining time derivative from energy and space derivative from momentum.

The "stuff" of the wave equation, what is said to "wave," is called by the Greek letter "psi" (ψ). Schrödinger did not have a physical interpretation for this ψ. Nor did he have a way of using the wave function to make definite predictions about observations of particles, though he assumed one would eventually surface. However, his colleagues pointed out that the square of the wave magnitude at a particular point yields the probability density for finding a particle there. That is, when used to calculate observable quantities such as position x or momentum p, Schrödinger's wave function yields probability distributions, with the probability density for finding the particle at a particular position proportional to the square of the magnitude of the wave function at that point.

There is a natural qualitative way to discuss the wave function: A pure wave with definite wavelength and direction is by definition infinite in extent—it oscillates in the same way no matter how far forward or backward we look at it. To make a localized wave in which the magnitude falls rapidly as we go away from its center, we need a *superposition* of pure waves, that is, we need to add waves of different wavelength and direction with suitable coefficients, so that they all add constructively in their contribution to the magnitude at the center in position space, but interfere destructively the more one looks away from that center. Such an array is called a *wave packet*.

We can choose coefficients of different pure waves which themselves have a maximum magnitude for a given wavelength and direction, and fall away as we go to different pure waves. These coefficients then form the wave function in "wavelength space," and when we add the de Broglie relation between momentum and

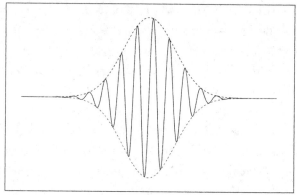

A Gaussian wave packet, or wave with a range of wavelengths that is localized in a region of space rather than stretching out to infinity in either direction, as a wave with just one wavelength must do. The dashed line in the figure is the envelope of the wave, which has a very smooth variation peaked at the center of the wave packet. This wave packet may be considered as a superposition of plane waves that add constructively in the region where the wave packet is large, but cancel among themselves far from the region of the wave packet.

wavelength, $p = h/\lambda$, this gives the wave function in "momentum space."

Schrödinger's approach is very satisfying, because it correctly gives the energies of different electron waves bound to an atomic nucleus to make an atom. At the same time it is disturbing, because electrons from the beginning had been accepted as particles; how could they also be described as waves? Furthermore, though Schrödinger himself never fully accepted the probabilistic interpretation, most others did, and it seems to confirm, once again, the fundamental role of probability in the structure of the world: The square of the magnitude of the wave function determines the probability density for finding an electron at any particular point.

Schrödinger used the familiar mathematics of waves to deduce the allowed frequencies of light emitted from an atom as it de-excites. Naturally enough, his formulation was called "wave mechanics."

HEISENBERG'S APPROACH

Heisenberg, working a few months earlier than Schrödinger, used an entirely different mathematics for the same end, which became known as "matrix mechanics." He started with the idea that one should not even try to describe classical orbits of an electron in an atom—to give a physical interpretation of what was happening. Instead, he felt, one should focus on the observable quantities, in particular the position and momentum of the electron, and look for the rules that govern how they evolve in time. Heisenberg, like Schrödinger, used differential equations—not for structuring the wave function, however, but for relating the observables x and p. Furthermore, those equations are exactly the same as Newton found nearly three centuries earlier! What, then, had changed—what made this a quantum version? The answer is that now the observables are not just numbers, which obey the commutative law of arithmetic, $pq - qp = 0$, but instead they are *matrices* or *operators*, obeying the relation $pq - qp = \hbar/i$. Here i is the imaginary unit, obeying the condition $i^2 = -1$. The symbol $\hbar$ is defined by $\hbar = h/2\pi$. As a result of this apparently slight change, the allowed energies of electrons in an atom no longer are continuous, but rather a discrete set of "quantum" energy levels. Further, it no longer is possible to predict exactly what position and momentum will be in the future from knowing, as best possible, their values at a given time.

CONSOLIDATION

Schrödinger, and then others, soon showed—despite the vast apparent differences in approach—that the two formulations were mathematically equivalent.

The key to understanding how there can be a straightforward relationship between the Heisenberg and Schrödinger approaches

lies in the *commutation relation* mentioned earlier, $pq - qp = \hbar/i$:
Schrödinger had found a differential equation for the wave function ψ, analogous to the Maxwell equations for light. Such an
equation relates *derivatives* of functions with respect to time and
position. A derivative with respect to x, df/dx, of a function at a
point x simply gives the rate at which that function is changing, per
unit change in x. In Schrödinger's equation a natural interpretation
for the derivative with respect to position in the x direction, for
example, describes it in terms of particle momentum in the x
direction: $d\psi/dx = ip\psi/\hbar$. Standard rules about differentiation then
give the Heisenberg commutation relation.

The approaches—taken separately and together—revolutionized
the theory of radiation. Suppose an atom is in a particular state
and makes a transition to another, lower-energy state by emitting
a photon. By the principle of conservation of energy, the difference in energy of the two states must be equal to the energy of
the photon, which in turn is related to the photon frequency by
the original Planck relation $E = h\nu$. Using either Schrödinger's
or Heisenberg's consistent rules, we can calculate systematically,
without need of the many ingenious guesses of Bohr's and Sommerfeld's, the energy difference of two hydrogen atomic states.

In addition, quantum mechanics can now be applied to more
complicated systems, such as atoms with more than one electron;
even though the calculations get more complicated, the answers
still are unique and well-defined. The old theory was helpless for
these systems. Heisenberg's and Schrödinger's work thus began an
accumulation of evidence for the power, consistency, and completeness of quantum mechanics that has continued to the present day.

Chapter Seven

Uncertainty

"BUT YOU CAN'T GO THROUGH LIFE APPLYING HEISENBERG'S UNCERTAINTY PRINCIPLE TO EVERYTHING."

"On or about September 1927," writes philosopher Ray Monk, "the physical world changed." Monk, consciously adapting a famous pronouncement by Virginia Woolf, is dead-on.[1] For hundreds of years, nature was cleanly separable from the human beings who studied it, and its parts all had a distinct identity and location, obeyed the same laws across all scales, and behaved predictably. During its first quarter-century, the quantum had led some physicists to question, reluctantly, certain of these assumptions. But the appearance of the uncertainty principle in 1927—published by Heisenberg in Germany in May, propounded by Bohr at a conference in Como, Italy, in September, and inaugurating a lifelong debate between Bohr and Einstein in Brussels at the Fifth Solvay Conference in October—was sensational, in-your-face proof these assumptions were false. It was "Exhibit A" in the case that a bridge cannot be built between the Newtonian and the quantum worlds. Nature was evidently a far stranger place than anyone thought.

Physicists and philosophers are still debating how to put finishing touches on the new picture of nature heralded by the uncertainty principle. Popular culture, however, began discovering and exploiting implications of the uncertainty principle shortly after its appearance in 1927. Until then, quantum physics was little discussed in popular culture. Newspapers and magazines treated it as something of interest because it excited physicists, but too complicated to explain. Even philosophers, who are professionally interested in scientific developments that alter our view of the world, had not seen quantum physics as posing particularly interesting or significant philosophical problems.[2] Soon after 1927, the uncertainty principle became the best known and most widely applied of all terms begotten by quantum mechanics.

It even sparked an entire genre of jokes. In one, a police officer pulls over Heisenberg for speeding and asks, incredulously, "Do you know how fast you were going?" Heisenberg replies confidently, "No, but I know exactly where I am!" In another version, the police officer pulls Heisenberg over and says, "Did you know that you were going 90 miles an hour?" Heisenberg says, "Thanks. Now I'm lost." In still another variant, Heisenberg's lover complains, "You've got great speed, but your position is off!"

Everyone gets these jokes because everyone knows Heisenberg's principle involves a trade-off of knowledge between where a thing is and how it moves, and the more you know of one, the less you know of the other. The jokes are a little sloppy because instead of the technically correct term "momentum," which is a vector quantity involving speed and direction, they mention only speed.

The comedienne Emily Levine once tried a more accurate variation in a show: "Why did the physicist have to take the bus to work? Because he knew the exact momentum of his car keys!" In previews, she found that nobody but physicists got it (if he knows the keys' momentum he doesn't know where they are) and had to drop the joke. The word "momentum" made most people think too hard to laugh.

Other jokes involving the uncertainty principle are aimed at those who get the jargon:

Q. Why won't Heisenberg's operators live in the suburbs?
A. They won't commute!

This joke relies on the fact that, in quantum theory, position and momentum are not technically variables (stand-ins for specific values of a quantity) but operators (functions that act on states)—and that these operators, as mentioned last chapter, are *non*commutative.

The uncertainty principle appears in panel cartoons:

It even made its way onto a wine label!

The uncertainty principle has also been applied—seriously, not tongue-in-cheek or ironically—to acting, anthropology, friendship, journalism, love, philosophy, politics, psychology, relationships, religion, self-help, social behavior, and more. Some philosophers argue that the uncertainty principle clinches the case for philosophical idealism, while some religious authors claim it opens the door for scientific recognition of free will. In *The Ascent of Man*, the Polish-British scientist Jacob Bronowski calls the uncertainty principle the Principle of Tolerance, while in *Newton's Football: The Science Behind America's Game*, Allen St. John and Ainissa G. Ramirez use it to explain the zone blitz. For still others it seems to have spiritual significance. Here is a conversation, published in *American Theatre*, between the theater director Anne Bogart and Kristin Linklater, the noted vocal coach.

LINKLATER: Some thinker has said that the greatest spiritual level is insecurity.
BOGART: Heisenberg proved that. Mathematically.
LINKLATER: There you are.

Bogart and Linklater seem to think uncertainty means insecurity. Does it? Does Heisenberg's principle have *any* meaning outside physics? Many think not. Craig Callender, a philosopher at the University of California, San Diego, published an article in the *New York Times* entitled "Nothing to See Here: Demoting the Uncertainty Principle." He wrote, "Let's put an end to the misuse of quantum physics to validate outlandish metaphysical claims."[3] And the following is a conversation in John Updike's novel *Roger's Version*:

"Until the observation is made, it's a ghost. According to Heisenberg's uncertainty principle—"
My blood was up . . . I told Dale, "If there's one thing that makes me intellectually indignant around here it's the constant harping of calf-eyed students on quantum mechanics and the Heisenberg principle as proof of that hoary old philosophical monstrosity Idealism."[4]

But how and why did the uncertainty principle become pervasive and polymorphic?

Arthur Eddington

Why the uncertainty principle became so widely discussed so shortly after its discovery had to do with a lucky turn of events involving the British astronomer Arthur Eddington. A prominent and articulate astronomer, Eddington was the leading communicator of physics and astronomy in Britain between the world wars. Of Quaker background, he was a conciliator, and often sought to reconcile warring factions. After the First World War, for instance, he championed the cause of German scientists, including Einstein, who were being openly spurned by English scientists. In 1926, Eddington was invited to deliver the Gifford Lectures, the most prominent lecture series worldwide on religion, science, and philosophy, held in Edinburgh, the following year. He decided to address "the philosophical outcome of the great changes of scientific thought that have recently come about."[5]

Eddington began writing his lectures in mid-1926, a year after Heisenberg developed matrix mechanics and while Schrödinger was publishing papers on wave mechanics. He delivered the first lecture, "The Failure of Classical Physics," on January 21, 1927. Our entire conception of the physical world has been shaken to its core by quantum mechanics, he told those who packed the Natural History classroom of Edinburgh University. Shortly after Eddington delivered his final lecture in March, Heisenberg installed the capstone on the quantum revolution by announcing the uncertainty principle. Eddington spent the remainder of 1927 revising the lectures for publication, which now included the first clear explanation of the uncertainty principle for outsiders. The lectures—enthusiastically received by the public—were published in 1928 as *The Nature of the Physical World*.

Arthur Eddington (1882–1944).

Here's the gist of the uncertainty principle, Eddington wrote: "a particle may have position or it may have velocity but it cannot in any exact sense have both."[6] He described in detail how measuring one property meant that the other becomes imprecise. The problem is no fluke, but "a cunningly arranged plot—a plot to prevent you from seeing something that does not exist, viz. the locality of the electron within the atom. . . . an association of exact position with exact momentum can never be discovered by us *because there is no such thing in Nature.*" The uncertainty principle means we need a "new epistemology." "It reminds us once again that the world of physics is a world contemplated from within surveyed by appliances which are part of it and subject to its laws. What the world might be deemed like if probed in some supernatural manner by appliances not furnished by itself we do not profess to know." The quantum teaches us that science has been aiming at a "false ideal of a complete description of the world."

This has two remarkable philosophical implications, Eddington said. The first has to do with free will. The Newtonian world had settled this conflict, from the scientific perspective at least, in favor of determinism: human beings are simply machines, like little clocks, whose operations were as fully determined by forces and motions as anything else in the universe. But the discovery of uncertainty shattered this, opening the door to more traditional religious ideas. A narrow-minded, rational person, Eddington said, might even conclude that "religion first became possible for a reasonable scientific man about the year 1927."[7]

The second implication was that the scientific world is only part of THE world: "We recognize a spiritual world alongside the physical world." As a result, "the physicist now regards his own external world in a way which I can only describe as more mystical, though not less exact and practical, than that which prevailed some years ago."[8] By setting limits on what science could know, Eddington said, quantum mechanics implied the validity of other kinds of knowledge, including mysticism. He saw quantum mechanics as giving birth to a kind of cultural commune, a human rainbow coalition where everyday humans, religious believers, and quantum physicists could stand arm in arm in a common spiritual quest.

Bridgman

Some scientists agreed with Eddington's thoughts, while others disagreed. Most scientists in the latter category politely ignored him, viewing him once again as playing the conciliator, using the prestigious pulpit of the Gifford Lectures to propose the possibility of a deep harmony between science and religion.

The Harvard physicist Percy Bridgman could not ignore Eddington. Several years later he wrote, "I still cannot think of his book *The Nature of the Physical World* . . . without bridling at the sheer bunk of a good deal of it. I regard Eddington as the supreme example of a crystal clear expositor of ideas as murky as mud." For a few months after Eddington's book appeared, the future Nobel laureate fulminated. Then, just before Thanksgiving 1928, Bridgman did something rash. He put down his instruments, brushed aside his scientific literature, and set out to enter the vulgar world of writing a magazine article. He sent a letter to *Harper's*, a magazine that occasionally ran articles about recent scientific developments.

Bridgman (1882–1961) had never written a magazine article before. Little had prepared him for it. He was born in Cambridge, Massachusetts, and spent most of his life at Harvard, entering as

Percy Bridgman (1882–1961).

an undergraduate in 1900 and remaining there as a graduate student, instructor, and professor. He was a shy, diligent, and careful man who strictly regulated both his exterior and inner self. Neither smoker nor drinker, he attended gym classes regularly, climbed mountains, and valued precision. He prided himself on leading a calm, orderly life adjusting gauges and cleaning equipment. He cared nothing for power or wealth and scorned those who sought them; for him the only thing worth pursuing was the slow and methodical process of increasing human knowledge. Years later, in fall 1946, when an excited student (Gerald Holton, now an eminent science historian at Harvard) burst into Bridgman's lab, defying instructions, with the news that someone was on the phone saying Bridgman had won the Nobel Prize, Bridgman did not even look up. Staring at his instruments, he said quietly, "Tell them—I'll believe it—when I see it."[9]

In 1927, Bridgman published *The Logic of Modern Physics*, an unusual book for a scientist. It did not try to popularize science, nor to smooth over its bizarre features to comfort the reader. Instead Bridgman candidly admitted the "disquietude" that relativity, with its strange implications for space and time, stirred in his colleagues. He shared this unease with readers, believing that you only understood the science if you understood this unease. The best we can do, he wrote in the first articulation of the philosophical position called operationalism, is treat certain scientific concepts as "nothing more than a set of operations," such as fixing the procedures for length.

Bridgman wrote little of quantum theory in the *Logic*. However, once he heard about the uncertainty principle, his concerns grew. The new quantum mechanics was highly rational, but not Newtonian. It shattered most assumptions on which he had been raised. These included Lord Kelvin's remark that "If I can make a mechanical model, I can understand it. As long as I cannot make a mechanical model all the way through, I cannot understand it." That was the implicit spirit of science, Bridgman realized. That spirit was now *gone*. What now? Eddington's touchy-feely, reality-is-everywhere stuff? The issue threw Bridgman into "a cognitive and moral crisis" for which he felt neither intellectually nor emotionally prepared.[10] He felt sure that the public would have an even more difficult time.

He wrote to a friend that if Heisenberg's principle holds up, it will mean "the biggest revolution in mental outlook since at least the time of Newton, much bigger than Einstein, for example." The world's structure disintegrates to meaninglessness at smaller and smaller realms. Wait until the popular culture gets wind of this, he mused—the result will be disastrous. The free-willers, the "pure chance" atheists, and the fans of vitalism in biology will all claim victory.[11]

Along came Eddington's book, which seemed to realize his worst nightmare—that people would say quantum mechanics proved the existence of God, and demonstrated the truth of mystic experience and new kinds of reality. No, no, no, no, no! Its implications were *not* comfortable and uplifting; they were deeply disturbing for they undermined what we thought reality was. He felt strongly that the public needed to understand that.

"Dear editors," he wrote to *Harper's*, "I am enclosing a manuscript dealing with some of the general and evolutionary implications of the recent discoveries in physics." He outlined his expertise. Then he continued, "I would particularly like to get this before the popular audience of *Harper's* because I believe that the consequences of the new discoveries are so important for everyone that all of us, sooner or later, will have to make considerable readjustments to meet the situation." In the manuscript, Bridgman wrote that quantum theory would soon exert a cultural impact greater than evolution and relativity, greater even than Newton and his work. Thanks to the uncertainty principle, discovered only a few months previously, quantum theory appeared to put a limit on what could be known—which, Bridgman predicted, would "let loose a veritable intellectual spree of licentious and debauched thinking." He continued:

> This imagined beyond, which the scientist has proved he cannot penetrate, will become the playground of the imagination of every mystic and dreamer. The existence of such a domain will be made the basis of an orgy of rationalizing. It will be made the substance of the soul; the spirits of the dead will populate it; God will lurk in its shadows; the principle of vital processes will have its seat here; and it will be the medium of telepathic communication. One group will find in the failure of the physical law of cause and effect the solution of the age-long problem of the freedom of the will; and on the other hand the atheist will find the justification of his contention that chance rules the universe.

Bridgman wrote to the editor, "I hope I am not irretrievably wrecking the Magazine." But he really didn't care; he wanted to set things straight. Those who promoted nonsense about the quantum did not know what they were talking about. The implications of the new physics were far more bizarre and unsettling than anything he'd heard so far.

An associate editor at *Harper's* returned his letter, and accepted Bridgman's article. *Harper's* published "The New Vision of Science" in March 1929.

"The thesis of this article," Bridgman wrote in his introduction, "is that the age of Newton is now coming to a close, and that recent scientific discoveries have in store an even greater revolution in our entire outlook than the revolution effected by the discovery of universal gravitation by Newton." These discoveries, in the field of quantum physics, show that we were mistaken to think that nature is fundamentally understandable and law-governed. The uncertainty principle, Bridgman wrote, was as "fraught with the possibility of greater change in mental outlook than was ever packed into an equal number of words." He explained that the impossibility of measuring exactly both the position and momentum of an electron means that electrons do not *have* position and momentum; in accord with his operationalist philosophy, he wrote that "no meaning at all can be attached to a physical concept which cannot ultimately be described in terms of some sort of measurement." More pointedly:

> To carry the paradox one step farther, by choosing whether I shall measure the position or velocity of the electron I thereby determine whether the electron has position or velocity. The physical properties of the electron are not absolutely inherent in it, but involve also the choice of the observer.

This principle, Bridgman went on, "probably governs every known type of action between different parts of our physical universe." This is "enormously upsetting" because it undermines our ideas of cause and effect. Wherever the atomic physicist looks on the atomic or electronic level, "he finds things acting in a way for which he can assign no cause, for which he never can assign a cause, and for which the concept of cause has no meaning." The reason, Bridgman said, "is not that the future is not determined in terms of a complete description of the present, but that in the nature of things the present cannot be completely described." Some German physicists have concluded that the world is governed by chance, but that is incorrect; quantum mechanics gives a definiteness and inevitableness to the subatomic world, but not the conventional kind.

The implications are staggering. "The physicist is here brought to the end of his domain." Until now, physicists have forayed ever deeper into nature, finding ever-finer structures. Now they come to a hitherto unimagined possibility:

> The physicist thus finds himself in a world from which the bottom has dropped clean out; as he penetrates deeper and deeper it eludes him and fades away by the highly unsportsmanlike device of just becoming meaningless. . . . A bound is thus forever set to the curiosity of the physicist. . . . The world is not a world of reason, understandable by the intellect of man, but as we penetrate ever deeper, the very law of cause and effect, which we had thought to be a formula to which we could force God Himself to subscribe, ceases to have meaning. The world is not intrinsically reasonable or understandable; it acquires these properties in ever-increasing degree as we ascend from the realm of the very little to the realm of everyday things; here we may eventually hope for an understanding sufficiently good for all practical purposes, but no more. The thesis that this is the structure of the world was not reached by armchair meditation, but it is the interpretation of direct experiment.

Some people are sure to draw rash assumptions, Bridgman continues, attacking Eddington without mentioning him:

> This will come from the refusal to take at its true value the statement that it is meaningless to penetrate much deeper than the electron, and will have the thesis that there is really a domain beyond, only that man with his present limitations is not fitted to enter this domain. . . . The man in the street will, therefore, twist the statement that the scientist has come to the end of meaning into the statement that the scientist has penetrated as far as he can with the tools at his command, and that there is something beyond the ken of the scientist.

This, then, is the imagined beyond which will become the playground of mystics and dreamers, lead to "an orgy of rationalizing," and comfort both atheists and religious believers. But Bridgman saw a silver lining. Eventually, he hoped, human beings can develop "new means of education" to inculcate into people the "habits of thought" required to reshape the thinking we use in "the limited situations of everyday life." The end result will be positive:

> [S]ince thought will conform to reality, understanding and conquest of the world about us will proceed at an accelerated pace. I venture to think that there will also eventually be a favorable effect on man's character; the mean man will react with pessimism, but a certain courageous nobility is needed to look a situation like this in the face. And in the end, when man has fully partaken of the fruit of the tree of knowledge, there will be this difference between the First Eden and the last, that man will not become as a god, but will remain forever humble.

Bridgman was prophetic. Humanity is still struggling with the task of getting humble, but he was right about the dreaming and the rationalizing.

Bridgman, his biographer wrote, was a "scientific puritan." He set demanding principles for himself that extended to the farthest corners of his life. Literally. A firm believer in assisted suicide for those with terminal illnesses, he thought that human beings were not meant to live without potential for dignity. When he contracted cancer, he lived his life as long as he could, then in 1961 sought a doctor to help him end it. When he failed to find one, he acted alone. "It isn't decent for Society to make a man do this thing himself. Probably this is the last day I will be able to do it myself," he wrote, minutes before raising a pistol to his head.

Eddington and Bridgman saw the implications of quantum mechanics in dramatically different ways. Eddington was the conciliator, willing to acknowledge the truth in everything, Bridgman was more careful, unwilling to venture beyond what he knew. Eddington saw quantum mechanics as pointing to a broader sense of reality than the traditional one, Bridgman to a narrower sense.

Eddington found the implications of quantum mechanics comforting, Bridgman disturbing. Each was in possession of different parts of the truth.

Other intellectual heavyweights, such as the philosophers John Dewey and Bertrand Russell, and scientists Arthur Compton and James Jeans (*The Mysterious Universe*, 1930), soon also weighed in. Alerted by the dispute between Eddington and Bridgman, Kaempffert, the *New York Times* reporter who had heard Heisenberg's first public lecture on the uncertainty principle, now saw the implications.[12] In a full-paged *New York Times* story published in 1931, he invoked Eddington and Bridgman and tried to digest the implications of the uncertainty principle. Atop the page were photographs of Einstein, Eddington, de Broglie, Schrödinger, and Heisenberg; further down were two images from Jeans's book *The Mysterious Universe*. The banner headline and subhead:

HOW TO EXPLAIN THE UNIVERSE? SCIENCE IN A QUANDARY
One After Another the Theories Put Out by the Scientists Have Been Exploded and Now Science in Its Uncertainty Has Been Forced to Become Idealistic and to Drop the Idea of a Mechanical Universe

What we thought we knew about reality, Kaempffert wrote, was based on ignorance; our Newtonian habits were bad habits. The fact that measurements depend on the choice of the observer "is much more than mere philosophical quibbling," he continued. "We are trying to explain reality, the universe, the objects we see and feel. We must of necessity explain them in terms of electrons, the elements of which they are composed. We find ourselves baffled. . . . Hence we can never know anything definite about an atom and therefore about a world composed of atoms. Pure chance reigns. The concept of cause has no meaning in the atom." But Kaempffert's article was a journalistic expression of what Bridgman feared: "murky as mud."

Polymorphous Metaphor

The phrase "quantum leap" became a popular metaphor in the early 1920s because of several factors: the need for a word where existing words feel inadequate (in contexts involving discontinuities), the desire of creative speakers to be more precise, and the cultural cachet of scientific terms. In the late 1920s, the uncertainty principle followed the same path, but faster and with more polymorphic applications. As a technical physics principle, it applies only to certain measurement situations, those involving conjugate variables; still, only one such case was sufficient for it to become a metaphor or to have implications beyond science. Four of the most common and significant of these involved unpindownability, the observer effect, emancipation of the arts and humanities, and the existence of God.

Unpindownability

The uncertainty principle provides an image that seems to ratify our experience of the world's randomness, lack of control, and even irrationality. A few days after the stock market crash of October 29, 1929, the *New York Times* invoked Heisenberg, tongue-in-cheek, as a partial explanation for the sudden disruption of the usual link between stock price and earnings: "Heisenberg had proved that the law of cause and effect no longer operates."[13] In 1936, the *New York Times* sportswriter John Kieran cheekily used the uncertainty principle, as expounded by Bridgman, to explain why the erratic New York Giants, recently in fifth place, still stood a chance. "His [Bridgman's] report of the rebound angles of billiard balls was all right in its way, but the rebound of baseballs off Giants' bats for the past six weeks has been a more convincing demonstration of the Heisenberg Principle of Uncertainty."[14] (The Giants ended up in the World Series that year—but lost to the New York Yankees.)

Sportswriters have made use of the uncertainty principle image ever since.

Others thought that the unpindownability implied by the uncertainty principle had a far deeper significance for pressing social issues. Religious thinkers in particular seized on Eddington's exposition. For centuries through the Newtonian Moment, scientists had them on the defensive by insisting that free will was an illusion. Here was a prominent scientist saying that science allowed free will after all, thus liberating mankind from the tyranny of atheistic determinism. "We are having our 'innings,'" wrote one religious commentator, using another baseball image, "and Babe Ruth, in the shape of Eddington, is at the bat to knock a few home runs."[15] More recently, the religious poet Christian Wiman, in his moving account of facing cancer, writes that "to feel enduring love like a stroke of pure luck" amid "the havoc of chance" makes God "the ultimate Uncertainty Principle."[16]

Observer Effect

That observing changes human behavior is a truth known informally to attentive human beings since ancient times and formally to contemporary psychologists, who call it "reactivity" or the "observer effect." The Newtonian world also understood that measuring things disturbs them, but assumed that the disturbance in principle can be made infinitely small and the measuring increasingly exact; the disturbance could also be anticipated and thus compensated for. The scientists could therefore "take themselves out" of their measurements.

In the quantum world, things are more complicated. How precisely you measure one of a pair of conjugate variables, such as position and momentum, affects the precision that you are able to obtain of the other. A crude way of putting this is to say that an experimenter "disturbs" a particle in observing or measuring it. This is wrong, for it implies a particle already equipped with position and momentum beforehand, which is merely jostled by the measurement. But in quantum mechanics position and momentum are undefined before a measurement, and whatever it is—it is odd to call something without position and momentum a "particle"— only acquires them when in contact with experimental equipment. Nevertheless, that common, shorthand description made the uncertainty principle accessible and available for a range of new metaphors in the human world.

This idea has inspired many cartoons about reciprocity: our experience that observation or inquiry changes what you observe.[17] If you press one thing you undo another. If you know x you can't

know *y*. The humor can go in many directions; for instance, it seems a scientific confirmation of something we already know, or of some absurd extension of what we know.

But alongside these playful and fanciful uses of the uncertainty principle's implications about observation we find serious ones that involve the nature of journalism, literature, and drama. In "How You Get That Story: Heisenberg's Uncertainty Principle and the Literature of the Vietnam War," the playwright and English professor Jon Tuttle notes that the uncertainty principle has been widely discussed by playwrights and journalists. The reason is the "social corollary" of the principle: "that the act of observing an event *changes the nature* of that event." This is for two reasons, Tuttle says: first, "the event immediately becomes relative to the observer," and second, that "observing the behavior of people who know they are being observed changes their behavior." Journalists after all, are professional jostlers. Tuttle continues:

Arthur Miller's *The Archbishop's Ceiling* and Tom Stoppard's *Hapgood* actually dramatize it [the uncertainty principle], showing theater audiences their own susceptibility to inaccurate perception and thereby implicating them as unwitting participants in the dramas they thought they saw. In an editorial, George Will cited the Uncertainty principle as an argument against televising jury proceedings, and Anna Quindlen, among others, has applied it to television "reality" shows, which of course stop being real the moment the cameras start rolling. Michael Herr realized the same problem in Vietnam when, for instance, soldiers would reposition cadavers for photogenic effect, or when the reporters' very presence would cause soldiers to amp up their courage and "promise the most awful kind of engagement," sometimes with tragic results.[18]

Still more seriously, the eminent literary critic George Steiner has used the uncertainty principle to describe the process of literary criticism, which transforms the object interpreted and delivers it differently to the generation that follows. "The two principles of indeterminacy and of complementarity, as stated in particle physics," he writes, "are at the very heart of all interpretative and all critical proceedings and acts of speech in literature and the arts."[19]

Emancipation of the Arts and Humanities

Just as many religious thinkers felt on the defensive during the Newtonian world, so did many practitioners of the arts and humanities. To them, the uncertainty principle's establishment of limits to measurement and knowledge seemed emancipating. J.W.N. Sullivan, reviewing Eddington's book *New Pathways in Science*, a follow-up to *The Nature of the Physical World*, voiced a widely felt feeling of relief:

> Everybody but a few die-hards is pleased with the new modesty that science has developed. It is no longer necessary for mystics and artists, in mere self-defense, to ignore the scientific outlook. Science has become aware that its knowledge is a particularly restricted kind of knowledge. Also, it has no exclusive claim to truth. It is admitted that there may be kinds of knowledge which are valid, even though they have not been, and cannot be, reached by scientific methods.[20]

The uncertainty principle's message seemed to be that science had no particular privileged perspective on truth, expanding the value of creativity in the arts and humanities.[21] Listen to the excitement of the Austrian Mexican artist Wolfgang Paalen:

> The capital crisis that science is traversing at this moment would perhaps justify the prediction of a new order in which science will no longer pretend to a truth more absolute than that of poetry, in which science will understand the value of art as complementary to her own. . . . If, in other words, in the highest possible precision of observation the perfect distinction between instrument and matter of experience becomes uncer-

tain, are we not permitted to conclude that physics is on the point of abandoning its pretension of offering us a purely quantitative and yet satisfying interpretation?[22]

The German philosopher Martin Heidegger, however, found such sentiments absurd and even dangerous. For him and others, the arts and humanities need no special defense. "The fact that we today, in all seriousness, discern in the results and the viewpoint of atomic physics possibilities of demonstrating human freedom and of establishing a new value theory," he wrote, is a sign of the harmful grip that "technological ideas" have on human self-understanding, and a complete misunderstanding of the nature of freedom.[23]

The Existence of God

Some people went even further, claiming that the uncertainty principle had implications for God's existence. These included scientists. Arthur Holly Compton, discoverer of the Compton effect, said that the uncertainty principle implied "the possibility of mind acting on matter," and a new role for human thought in the universe.[24] Does the uncertainty principle imply that there must be a nonhuman Knower? Yes, Compton answered. Quantum mechanics points to "the existence of a supreme being guiding the affairs of the universe, in which man probably represents the highest order of intelligence." He said, "I believe that the very existence of the amazing world of the atom points to a purposeful creation, to the idea that there is a God and an intelligent purpose back of everything."[25]

Religious writers were ecstatic. The *Methodist Review* wrote: "Science, as transformed by the new physics, taught by such physicists as Compton, who has won the Nobel prize, and Heisenberg, one of the masters of the Quantum theory, has transfigured Evolution from a process of mere chance to a creative method of an unseen Personality, one who has in this developing process a definite aim toward which he is moving."[26] In an article headlined

"New Theory of Universe Said to Place Intelligent Purpose Behind Nature," the *Christian Science Monitor* declared, "Discovery of the new 'uncertainty principle' of the universe has rendered untenable the theory that men are merely automatons and provides strong evidence of the existence of 'a directive intelligence,' according to Dr. Arthur H. Compton, Nobel Prize winner in physics."[27]

In 1929, the *Methodist Review* ran several articles on the impact of quantum physics on religious belief. Eddington, the *Review* declared, "reveals the coming of a new science." In his new epistemology "no contradiction can ever be allowed between science and religion." The quantum view "is making its masters as imaginative as any spiritual mystic." And "Science evidently does have at present its separate watertight compartments of thought, but religion must ever fill the entire world of thought, feeling and will. Indeed the quantum physicists of to-day are producing a view of nature so indeterminate that it could fairly be called the *super*-natural." Heisenberg, it declared, may well be the greatest living scientist to influence metaphysics and religion. "[W]e are entering a view of the physical world that brings it into closer harmony with that larger realm of perfect freedom in which Christianity lives, the kingdom of God."[28]

Others were appalled at what they thought a misappropriation of quantum mechanics by religious authorities and other enemies of social progress. The radical African American labor organizer Ernest Rice McKinney said of Eddington, "He gave great consolation to the preachers when he seized on Heisenberg's Uncertainty Principle, giving the doctrine of causality and determinism a black eye and giving these religionists an opportunity to rehabilitate free will." McKinney, a Marxist and atheist, liked Jeans's account better. Jeans speaks of God as an architect but at least also calls Him a mathematician: "I prefer a God who is a pure mathematician to one who is a major general, a corporation president, a grand wizard, and engineer or a theologian."[29]

Heisenberg did not introduce uncertainty into the human world, nor did humanity need his principle in order to see it in human life. The principle did, however, reveal that several Newtonian assumptions about reality were incorrect. To some extent, the quantum's impact on artists, writers, and philosophers was that it helped free them from these Newtonian-inspired misconceptions. A year or two after the discovery of the uncertainty principle in 1927, for instance, the writer and poet D. H. Lawrence penned the following poem fragment:

> I like relativity and quantum theories
> Because I don't understand them.
> And they make me feel as if space shifted about like a swan
>   that can't settle,
> Refusing to sit still and be measured;
> And as if the atom were an impulsive thing
> Always changing its mind.[30]

The uncertainty principle, as we have seen, only says that certain pairs of variables cannot be measured simultaneously—and far from making the atom "impulsive" it is responsible for its rigidity. Still, the fragment is revealing in that it expressed the fact that quantum theories captured better than Newtonian ones Lawrence's intuitions about the world. He feels more at home—freer, liberated from being told that he's determined, OK with being something that doesn't settle. The quantum world does not seem strange to him because his intuition tells him that our world is stranger than Newton said it was.

Lawrence's playfully negative remarks may suggest that his attraction is superficial: he likes relativity and quantum theories because they connect better with his experiences of the world as quixotic and immeasurable. The quantum world seems to have "space" for feelings that were suppressed in the Newtonian world. Paalen expressed a similar sentiment in 1942, as we have seen, when he wrote excitedly that quantum mechanics heralds "a new order in which science will no longer pretend to a truth more absolute than that of poetry," and that the outcome will be to legitimize the value of the humanities, and "science will understand the value of art as complementary to her own."[31]

In 1958, the New York University philosopher William Barrett (1913–1992) wrote that the principle "shows that there are essential limits to our ability to know and predict physical states of affairs, and opens up to us a glimpse of a nature that may at bottom be irrational and chaotic—at any rate, our knowledge of it is limited so that we cannot know this not to be the case. This finding marks an end to the old dream of physicists who, motivated by a thoroughly rational prejudice, thought that reality must be predictable through and through."[32] After citing a few more examples from modern science, Barrett continued, "What emerges from these separate strands of history is an image of man himself that bears a new, stark, more nearly naked, and more questionable aspect."[33] Not only existentialist philosophy but even science itself has forced man to confront his solitary and groundless condition in "a denudation, a stripping down, of this being who has now to confront himself at the center of all his horizons."

Such remarks suggest that the reason humanists embraced quantum mechanics was that they experienced the Newtonian world as a cold and constricting place in which they felt defensive and marginalized, so the news of the strangeness of the quantum domain came as a relief. But if this is the only reason humanists found developments of the quantum world liberating, it was surely their own doing, for they were relying far too seriously on science to begin with in understanding their own experience. Barrett's remarks reveal that the uncertainty principle even has a tragic air about it. In *Oedipus Rex*, the protagonist, seeking one destiny, inexorably brings about the opposite by his desire for control; in the subatomic world, when human beings seek to know through the powers of ever-more precise measurement, they inexorably introduce imprecision. That is a philosophical shocker.

The Slovene philosopher Slavoj Žižek has done much to draw out the philosophical implications of the uncertainty principle. It's not about inherent limitations in the observer or the instruments, Žižek recognizes. Rather, it indicates that an uncertainty is "inscribed into the thing itself," so that, say, an electron's very being is to be able to show itself as either particle or wave, but never at the same time. Žižek calls this "ontological cheating"—a thing that does not simply know what laws it is supposed to obey—and says that it opens up "the shadowy pseudo-being of pure potentialities." He then goes on to compare the genesis of determined, classical physical law out of quantum interactions and the genesis of symbolic order out of the activity of human beings.[34]

But scientifically, the uncertainty principle was like a new dawn. It explained a lot, including the origin of the randomness in quantum theory discussed in Chapter 4. There is no uncertainty at all in the time evolution of the wave packet. It spreads because there is in that packet an uncertainty between position and momentum, therefore a range of momenta, therefore a range of velocities, therefore a spreading. Still, the wave packet itself is completely well defined, and propagates over time with no uncertainty. The randomness arises not because of the wave itself, but because it describes only the probability of seeing a particle. If there were no uncertainty, there would be no randomness.

While the uncertainty principle's name suggests a squishiness, just the reverse is true; there is a hardness and definiteness in quantum physics that classical physics never could attain. According to classical physics, an electron circulating around a proton in the hydrogen atom would keep losing energy by radiation, eventually descending into the proton. But in quantum mechanics, an electron would have to have a huge amount of kinetic energy to descend into such a confined space, meaning that it would not stay there long, accounting for why the atom cannot collapse. For larger atoms with more electric charge on the nucleus and more electrons circulating around, the uncertainty principle energy is supplemented by the Pauli exclusion principle, so that such atoms also cannot collapse, and instead have very definite energies, always the same for a given atomic number. If we pile many of these atoms together, lower the temperature, and put on enough pressure, then they will assume a very rigid solid form. Thus uncertainty in one thing, the relation between position and momentum, leads to certainty in another, namely the structure of atoms, always the same for a given nucleus and a charge-balancing number of electrons. The uncertainty principle thus explains the "hardness" in matter.

In 1933, the Nobel prizes for the previous two years in physics were announced. The winner of the 1932 prize was Heisenberg; he was not yet thirty-two. Schrödinger and Dirac shared the 1933 prize. In December of 1933, the three of them stood on the same platform together at the Swedish Academy, for Heisenberg's prize had been delayed—a trio of architects of the new physical world.

The Uncertainty Principle

The uncertainty principle itself first appeared in letters between Heisenberg and Pauli. In fall 1926, Heisenberg insisted to Pauli that classical concepts such as position and momentum were inapplicable to the quantum world. Their conversation was joined by their colleague Pascual Jordan, who devised several ingenious thought experiments—what if you froze a microscope to absolute zero?—that would seem to make it possible to measure properties that Heisenberg and Pauli said could not be measured. Jordan's challenge stymied Heisenberg. Talking to Bohr usually helped clear his mind, and in February 1927 he visited Copenhagen. This time, however, the conversation went nowhere, the two grew snappish with each other, and Bohr went skiing, leaving Heisenberg to brood alone in Copenhagen. One night, on a solitary walk in the park behind Bohr's institute, Heisenberg saw the outlines of an answer. He sat down to write a letter to Pauli:

> One will always find that all thought experiments have this property: when a quantity p is pinned down to within an accuracy characterized by the average error p_1, then . . . q can only be given at the same time to within an accuracy characterized by the average error $q_1 \approx h/2\pi p_1$.[35]

This is the uncertainty principle. Like many equations, including $E = mc^2$ and Maxwell's equations, its first appearance is not in its now-famous form. Today we know it generally as $\Delta p \Delta q \geq \hbar/2$ (the new symbol on the right, introduced by Dirac the following year, standing for $h/2\pi$, proves simpler to use in many applications). Heisenberg was thrilled by the insight, and sent off a paper, "On the Visualizable Content of Quantum Theoretical Kinematics and Mechanics," published in May. The paper is remarkable in the way it relies on classical concepts and terms to reach its conclusion, would have been unthinkable without the large role indeterminism had come to play in physics, and carefully justified its counterintuitive conclusions using relativity as a model.[36] On his return, Bohr expressed unhappiness with the paper and pointed out errors, but Heisenberg went ahead and published an only slightly revised version.

In retrospect, the uncertainty principle was implied in the "mysterious equation" $pq - qp = Ih/2\pi i$, and even implicit in a way in Planck's quantization of phase space (the second of at least three major contributions Planck made to the new physics). Still, it was a conceptual breakthrough. Until that moment, Born, Pauli, Heisenberg, and Jordan had been discussing cases where one variable was exactly determined and the other not. Heisenberg now showed—contradicting in part his earlier insistence that classical terms did not apply—that these were limiting cases, with a range of possibilities in between.

Furthermore, the uncertainty principle does not apply to all properties of the microworld; mass and charge, for instance, have perfectly definable and precise values. It applies only to certain pairs of variables, called conjugate variables, and restores their visualizability (up to a point) at the cost of uncertainty; it says that a precise mathematical formula governed the margin of uncertainty in their values. Still, Heisenberg knew that more was at stake than domestic atomic physics; the uncertainty principle had "foreign" repercussions even for the nature of reality itself. "One might be led to the conjecture that under the perceptible statistical world there is hidden a 'real' world in which the causal law holds," Heisenberg wrote. "But it seems to us that such speculations, we emphasize explicitly, are fruitless and meaningless. Physics should only describe formally the relations of perceptions."[37]

The uncertainty in the principle, to use philosophers' words, is not just epistemological, having to do with what we know about nature, but ontological, having to do with nature itself. As Pais put it, the term "uncertainty principle" is therefore a misnomer: "The issue is not: what don't I know? But rather: what can't I know?" But, Pais added, it's too late to change the name now.[38]

After sending off the paper Heisenberg wrote to Jordan that he was excited to feel the "discontinuous ground under my feet," and Pauli recalled him saying something like "Morganröte einer Neuzeit"—the dawn of a new era.[39]

Chapter Eight

Reality Fractured: Cubism and Complementarity

"Fruitloopery" is the way *New Scientist* magazine refers to pretentious and erroneous use of scientific words. The term has a weird metaphoric origin. "Froot Loops" is the name of a breakfast cereal—made of small ring-shaped pieces in bright colors with fruitlike flavors—introduced by the Kellogg Company in 1966. For a time, "fruit loop" was US slang for male homosexual, or a neighborhood frequented by them, but the sexual connotation largely drained away and the word began to mean something lightweight, nutty, and pretentious.

In 2005 Mike Holderness, a freelance contributor to *New Scientist*, wrote of "professional dissidents" who are given the oxygen of publicity by those journalists who "divide all stories into precisely two sides that get equal space: too often the reality-based community versus fruitloops and/or special interests." Language needed a term like that, and Holderness's choice was inspired. "Fruitloopery" became the *New Scientist*'s generic word for advertisers' use of science either unverifiably or wildly out of context. Fruitloopery indicators in ads include the words *quanta, tachyons, vibrational energies,* or *restructured water,* especially in combination. In our class we extend the meaning further, and apply the term to any pretentious and erroneous use of scientific terms not only in advertising but also in self-help literature, amateur philosophy, and pseudoscience.

Physics inspires more fruitloopery than other fields, due to its cultural prestige. If you're a charlatan and you link something you are promoting to physics, you imply that it's deep and secure. Good for sales! It's not a mistake—not an error to be corrected. Charlatans don't make mistakes, only fruit loops. And money.

Within physics, quantum mechanics generates more fruit-loopery than other branches. But how do you tell the difference between fruitloopery and meaningful uses? It's much harder than it looks. Some of the quantum metaphors we've mentioned so far—Updike's, for example—are technically unsound but clearly meaningful.

The self-consciously humorous use of quantum concepts is not fruitloopery:

These clumsy uses of language are what we might call found quantum humor.

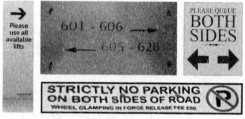

Assorted building and roadside signs,
examples of found quantum humor.

Can we tell fruitloopery by the credentials of who is speaking—who is serving up the cereal? Not necessarily. Here are some seemingly idiotic pronouncements about physics by actress Shirley MacLaine:

The new physics was teaching us that we were inextricably involved. That not only were we involved, but that such atomic structure might exist because of our consciousness . . . that the universe . . . might just be one giant, collective "thought."[1]

We can jeer at these remarks, that is, until we compare them to a passage from James Jeans's book *The Mysterious Universe*:

The stream of knowledge is heading towards a non-mechanical reality; the Universe begins to look more like a great thought than like a great machine. Mind no longer appears to be an accidental intruder into the realm of matter . . . we ought rather hail it as the creator and governor of the realm of matter.[2]

The language of the actress and the scientist is so similar that it's likely either that MacLaine read James Jeans's book, or read someone paraphrasing Jeans.

Fruitloopery is not subjective or arbitrary, and involves "magification." Something is fruitloopery when the words and images being invoked are treated as having special powers, like magic. Fruitloopery relies on borrowed authority, a presumed authority inherent in the words and images themselves. Advertisements for health remedies, lifestyle improvers, or intelligence boosters that claim to work by exploiting quantum powers are cases of fruitloopery. Numerology, the attribution of special powers to numbers, is also an example. Astrology, the claim that you can tell a person's character from the arbitrary groupings of stars—scattered billions of billions of miles away from each other and the earth—that certain planets are seen against from the position of the earth at the time that person emerges from the womb, is another.

But what's the danger of a little magic? Plenty. Think of the difference between someone (a friend or psychiatrist, say) proposing a description of your character and asking you whether it helps you understand yourself better, and someone (an astrologer, say) proposing that *same* description and telling you that you should believe it because the stars exert a power over you and have decreed that this is your character. The first case is supportive and enabling, the second is authoritarian and only works by exploiting your vulnerability.

Picturing Schizophrenia

The hardest cases of fruitloopery to detect involve "complementarity," Niels Bohr's 1927 neologism for a phenomenon of the subatomic world for which he could not find any good metaphors from the human world. It then became metaphorically applied to the human world.

Bohr's motivation was the following: By the 1920s, as we saw in Chapter 6, scientists realized that, from the perspective of the classical world at least, the quantum world was schizophrenic. Bragg tried to express this in his image of the log, the wave, and the ship, and in his saying that his colleagues use wave theory on Mondays, Wednesdays, and Fridays and particle theory on Tuesdays, Thursdays, and Saturdays, while J. J. Thomson referred to tigers and sharks. The quantum revolution of 1925–27 briefly made it seem like a solution was at hand. Heisenberg's matrix mechanics approach said: "Don't bother to look for a metaphor! Metaphor-seeking is exactly what gets you into trouble!" Schrödinger's wave mechanics approach said: "Waves are a fairly good metaphor after all!"[3] But when the uncertainty principle appeared in 1927 it was clear that neither approach was truly satisfactory. Schizophrenia was here to stay. Some "thing"—in a broad sense of the term that was hard to describe—could have completely irreconcilable properties. This was a philosophical conundrum within the Newtonian perspective, where things bore their properties on their backs, so to speak. What did reality and objectivity mean anymore?

Bohr practiced physics as if he were on a quest. His grail was to fully express the quantum world in a framework of ordinary language and classical concepts. "[I]n the end," as Michael Frayn has Bohr's character say in the play *Copenhagen*, "we have to be able to explain it all to Margrethe"—Bohr's wife and amanuensis who serves as the onstage stand-in for the ordinary (i.e., classically thinking) person. How do you explain waves and particles to your spouse, and yourself?

As Bohr was pondering this question, according to physicist and historian Arthur I. Miller, he found inspiration in the style of painting known as Cubism, which was pioneered by Pablo Picasso and Georges Braque about 1907 and gaining steam by 1909.[4] Wanting to escape the literalism of perspective painting, which duplicates the "photographic" view of an observer, Cubist artists sought to lay down a multiplicity of views from different directions on a single canvas, piled onto each other in a kind of collage. This makes "Cubist vision" more complete, in a sense, than normal vision. Each painting shows numerous incompatible views all at once, as if from nowhere and everywhere—but also shows how these incompatible views all fit together as views of the same thing, with each position of the spectator associated with a particular view. Bohr himself kept a Cubist painting by Jean Metzinger at home, and liked to use it as an illustration of complementarity.

Bohr therefore literally had in sight a pictorial analogue of the question he was addressing: how something like a quantum phenomenon could show itself simultaneously in different ways (as waves and particles) even though from any given "view" (experiment) one sees primarily the one or the other. Miller writes:

> Cubism directly helped Niels Bohr discover the principle of complementarity in quantum theory, which says that something can be a particle and a wave at the same time, but it will always be measured to be either one or the other. In analytic cubism, artists tried to represent a scene from all possible viewpoints on one canvas. An observer picks out one particular viewpoint. How you view the painting, that's the way it is. Bohr read the book by Jean Metzinger and Albert Gleizes on cubist theory,

Du Cubisme. It inspired him to postulate that the totality of an electron is both a particle and a wave, but when you observe it you pick out one particular viewpoint.[5]

In May 1927, a few weeks after learning of the uncertainty principle, Bohr began work on an essay, "Philosophical Foundations of Quantum Theory."[6] He struggled to turn it into publishable form but another project loomed: a talk he was supposed to give at a conference in September to commemorate the centennial of the death of Italian physicist Alessandro Volta, to be held in Volta's native town of Como. Bohr's philosophical essay went through several drafts and was never published, though he incorporated his

Woman on a Horse by Jean Metzinger, 1911–1912,
National Gallery of Denmark, Copenhagen.

thoughts into his Como speech. On September 16, 1927, Bohr gave his talk: "The Quantum Postulate and the Recent Development of Atomic Theory." It would introduce a new quantum metaphor into ordinary language: complementarity.

Bohr began slowly, repeating things he knew his audience had heard before. Our scientific equipment—the stuff handled by physicists in their laboratories—belongs to the classical world, and we've got no choice but to use classical physics to understand it. Usually, the things we study "may be observed without disturbing them appreciably." But quantum mechanics surprised us. "Any observation of atomic phenomena," Bohr said, "will involve an interaction with the agency of observation," and Heisenberg has just shown us that this interaction cannot be eliminated, is non-negligible, and of a certain quantifiable amount.

Then Bohr drops in a now-famous remark that merely lays out the logical consequence: "Accordingly, an independent reality in the ordinary physical sense can neither be ascribed to the phenomena nor to the agencies of observation."

Ordinary experience is unaffected by these implications because of the smallness of Planck's constant, just as ordinary experience is unaffected by relativity because of the huge speeds required before its effects can be noticed. Still, Bohr noted, "This situation has far-reaching consequences" for the nature of knowledge. In classical mechanics the definition of the state of a system requires "elimination of all external disturbances." It requires that we be outside the fishbowl, looking in without disrupting anything. But in quantum mechanics, "every observation introduces a new uncontrollable element," meaning that "an unambiguous definition of the state of the system is naturally no longer possible, and there can be no question of causality in the ordinary sense of the word." We can define it only if we don't observe it, and therefore if we don't really know it—and if we observe it, we screw up the possibility of definition. The nature of quantum theory thus forces us to regard the location of a particle in space-time, and its governance by causal laws—whose identity is a feature of classical mechanics—"as complementary but exclusive features of the description, symbolizing the idealization of observation and definition respectively." Quantum mechanics "presents us with the task of developing a 'complementarity' theory" of knowledge.

Complementarity was thus Bohr's name for this ontological disturbance, for the way quantum phenomena changed the kind of "things" they were depending upon how we accessed them. While the classical object is one thing that only looks different when accessed in different ways, the quantum object actually *becomes* different. It's as if, in the famous parable of the blind men who touch different parts of the elephant, compare notes, and are in complete disagreement, there *is* no elephant; the "quantum elephant" actually *becomes* a rope, pillar, fan, pipe, wall, and so forth, depending on how it's touched.

But at the end, Bohr announced a deeper lesson. "I hope that the idea of complementarity is suited to characterize the situation, which bears a deep-going analogy to the general difficulty in the formation of human ideas, inherent in the distinction between subject and object."

Bohr's Como talk, delivered in September 1927, was published in *Nature* the following April.[7] The editors, alarmed that Bohr and company viewed quantum phenomena as abstractions that only become concrete, defined, and observable when measured, took the highly unusual step of writing a preface. What's in the article, they wrote, "is far from satisfying the requirements of the layman who seeks to clothe his conceptions in figurative language," adding that "it is earnestly to be hoped that this is not their [the quantum physicists'] last word on the subject, and that they may yet be successful in expressing the quantum postulate in picturesque form."[8]

Pauli laughed when he read this, and satirically paraphrased the queasy editors' tone as follows: "We British physicists would be awfully pleased if in the future the points of view advocated in the following paper should turn out not to be true. Since, however, Mr Bohr is a nice man, such a pleasure would not be kind. Since moreover he is a famous physicist and more often right than wrong, there remains only a slight chance that our hopes will be fulfilled."[9]

Reception

Many of Bohr's colleagues, too, did not react well to the idea. It seemed to them to be philosophy—a term of opprobrium among physicists—unnecessary at best and confusing and obstructive at worst. Jammer summarizes the general reaction of the Como audience to Bohr's talk as that it "will not induce any of us to change his opinion about quantum mechanics."[10] Schrödinger thought that Bohr was attempting to sweep difficult philosophical problems under the rug, writing that "Bohr wants to 'complement away' all difficulties."[11] The philosopher Karl Popper wrote, "I do not doubt that there is an interesting intuitive idea behind Bohr's principle of complementarity. But neither he nor any other member of this school has been able to explain it."[12] Only a handful of other scientists were enthusiastic. J. Robert Oppenheimer remarked that complementarity was among the things that atomic physics has taught that "provide us with valid and relevant and greatly needed analogies to human problems lying outside the present domain of science."[13] On the other hand, Oppenheimer also remarked that we have better and more traditional sources of insight to help us understand irresolution in the human world than quantum mechanics: "Hamlet has said it better than Planck's constant."[14]

What were those supposedly relevant and needed analogies? Bohr was the first, and most diligent, at trying to identify them. He had a deep appreciation for philosophy and psychology, yet was frustrated by the inability of professionals in those fields to take what he thought was a serious interest in the new physics discoveries. He planned to write what some of his colleagues came to call "The Book," which would have been "a comprehensive presentation of complementarity and its implications" throughout all aspects of human life.[15] Historians have struggled to produce a clear picture of what might have been the contents of that book. In his biography of Bohr, Pais was stymied by his attempt to describe the significance of complementarity. He proceeded by writing a question and answer account of what generally happens when he attempts to teach complementarity to physicists. It includes the following exchange:

Q. All this about complementarity is very interesting. But of what use is it to me?
A. It will neither help you in your quantum mechanics calculations nor in setting up your experiment. In order to do physics you should not only assimilate and develop facts, however. In between you had better reflect on the meaning of what you are doing. In that respect Bohr's considerations are extremely significant. Don't you agree it ought to matter to you what, for example, a modern scientist means or should mean when he talks about "a phenomenon"? Note also that insights like these may serve to explain to interested laymen and to remind scientists what your profession is about.[16]

In the last line of his *Nature* article, Bohr spoke of complementarity as addressing a "general difficulty" in human ideas, hinting it had a wider application to other fields in which human beings are both actors and spectators. Noting this dual role of both acting and watching is also a feature of anthropology, biology, and psychology, Bohr tried to understand their problems in terms of complementarity. In biology, complementarity might describe the tension between proponents of mechanism (the view that life is a product of machinelike causality) and vitalism (life as the product of a special, individualized, animate force). In psychology, Bohr thought it might apply to the tension between first and third perspectives. He also explored its extension to other fields, including government, justice, and love. Had Bohr been successful, it would have been a dramatic instance where concepts developed in subatomic physics were applied literally—not metaphorically—in the human sphere. Yet his attempts were not well received.[17]

Many physicists were embarrassed by Bohr's attempts to extend complementarity outside physics. In 1998, the New York University physicist Alan Sokal mocked humanists for delving into physics to support their ideas in a way that seemed ignorant at best and zany at worst, crafting an article consisting mainly of silly and nonsensical remarks—many involving complementarity—made by scholars in the humanities about physics. Sokal succeeded in getting the article published in *Social Text*, a leading journal of the humanities, in an episode now known as Sokal's Hoax. In response, the science historian Mara Beller published an article in *Physics Today* entitled "The Sokal Hoax: At Whom Are We Laughing?" She cited remarks by Bohr—but also Born, Heisenberg, Jordan, and Pauli—to show that in this respect physicists could sometimes be as silly as humanists. One example:

> The thesis "light consists of particles" and the antithesis "light consists of waves" fought with one another until they were united in the synthesis of quantum mechanics. . . . Only why not apply it to the thesis Liberalism (or Capitalism), the antithesis Communism, and expect a synthesis, instead of a complete and permanent victory for the antithesis? There seems to be some inconsistency. But the idea of complementarity goes deeper. In fact, this thesis and antithesis represent two psychological motives and economic forces, both justified in themselves, but, in their extremes, mutually exclusive. . . . there must exist a relation between the latitudes of freedom *df* and of regulation

dr, of the type *dfdr* = *p*. . . . But what is the "political constant" *p*?
I must leave this to a future quantum theory of human affairs.[18]

This seeming piece of nonsense, Beller then revealed, was written by none other than Max Born, one of the founders of quantum mechanics and awardee of the 1954 Nobel Prize for the statistical interpretation of quantum mechanics. Beller's article is sobering—it showed how hard it is to distinguish between Bohr's attempts and those of people who are less rigorous, to describe the relevance of quantum terms and imagery in human culture.

The distinction became still harder to draw when the term became public knowledge. In 1933, Bohr visited the United States for his second time to address what is familiarly known as the "Triple A S," the American Association for the Advancement of Science, meeting in Chicago that June. He gave a public talk outlining his ideas about complementarity, whose message disappointed scientists but enchanted others.[19] Reporters found their own vivid metaphors to express Bohr's. "Jekyll-Hyde Mind Attributed to Man," announced the *New York Times*.[20] The article explained that Bohr was expanding the application of the uncertainty principle beyond the realm of atomic physics "to include man's entire relation to the world around him and to all processes of knowing and thinking." Bohr has discovered "an inherent essential duality in the nature of things, as they relate to man's ability to know them. The paradox of this duality lies in the fact that the Jekyll-Hyde nature of all things is essentially contradictory, with both aspects being true at different times, but with only one aspect being true at any one given time." In the 1886 Robert Louis Stevenson novella, the transformation of Dr. Jekyll into Mr. Hyde is (mostly) under Jekyll's own control—he drinks a potion to metamorphose into the amoral Hyde—while Bohr's idea was that the transformation between one incompatible atomic state and another depends on the measurer. The article continued:

In other words, the very process of knowing one aspect of nature makes it impossible for us to know the other aspect. We can know only one side of its nature at any one time. There is a definite discontinuity in all things partaking of existence and of knowledge, so that when one thing is true this very truth perforce makes another thing non-existent as far as any possible knowledge on our part is concerned. This contradictory duality is inescapable because it lies at the very heart of things. It is wrong, according to this theory, to say that there are either free will or determinism, causality or chance. Both are essential parts of one and the same reality, the convex and concave sides of the same sphere . . . when you are inside the sphere the sphere is concave and it is never possible to experience its convex aspect.

The article then said that Bohr and other quantum physicists had tried to kill the "monster of 'uncertainty,'" but kept finding it, "Mephistopheles-like in another form, mocking at them." The concept of complementarity was a concession to this irreducible fact.

The *Times* liked this image, and the next day started metaphorical riffs. An editorial entitled "Wedding of Opposites" wryly commented that Bohr's new theory of knowledge seemed "specifically designed to fit and define the ways and works of politicians," where "reconciling opposites seems to be the order of the day." The *Times* wrote, "Based on modern atomic physics, the Theory of Complementarity is the nearest thing we have seen to a scientific account of movements now going on in the political world at home and abroad. . . . Complementarity fairly dominates the scene in our municipal politics. Nothing is what it seems. Everything is in a daily flux under the eye of the observer." And referring to a Tammany Hall political action: "Should that manoeuvre succeed, and should Mr. Hylan suddenly become Republican and anti-Tammany, Heisenberg himself would admit that his scientific uncertainty had been far outstripped by practical politics."[21]

The way that "quantum leap" spread from the subatomic realm

to everyday language, we said in Chapter 3, models the path followed by other expressions and images in this book. They are "called" into metaphorical use by a combination of factors: the need for a word where existing words are inadequate, the desire to be more precise, and the cultural cachet of scientific terms. The phrase "quantum leap" was called into use for a range of metaphorical applications by the need to describe discontinuous gaps in an increasingly mechanized age. "Complementarity" was applied to incompatible approaches to the same subject matter, from art to religion.[22]

But don't we have these already? Ever since ancient times, both Eastern and Western civilizations have been familiar with symbols such as the mandorla—two overlapping circles within an almond-shaped whole—or the yin-yang sign—a wavy line within a circle separating two equal areas, each containing a dot—to capture the co-presence of opposites in human life and/or nature. The opposites may be good/evil, heaven/earth, life/death, spiritual/material, and so forth, with the whole embracing both parts. Bohr now produced a new word to capture the co-presence of contradictory ideas in his attempts to explain what was happening in quantum physics. It was embraced by journalists, writers, and scholars in many different fields. Bohr encouraged the connection to the ancient images when he chose a yin-yang symbol for his family herald. The word was especially attractive for several reasons. First, it had the cachet of a scientific term. Second, it said that the process of generating opposites was not impersonal or objective but brought about by human intervention. Human beings seemed to be important to the universe again. Leonard Shlain, a surgeon and author who wrote enthusiastic books about the meaning of art and science, declared that "Bohr's theory of complementarity . . . seems to border on the spiritual."[23] Others applied complementarity to politics and other areas, and declared it to come in many different forms, such as "horizontal" or "vertical."[24] Sometimes authors write of the complementarity of different aspects of one thing, while at other times they declare entirely different fields—science and religion, physics and art—to be complementary.[25]

At still other times, the complementarity metaphor is merely tongue-in-cheek. Take, for example, the Quantum Beer Theory website (now long gone, alas) once created by Kyle Wohlmut, a translator and dedicated beer fanatic who lives in the Netherlands. Wohlmut, who last studied physics in high school, explained to us that underlying his theory is the fact that "the essential experience of beer flavor arises from conflict" in a complementary way. In each beer, he claims, two sides of taste—hops and malt—struggle for supremacy. "These two sides wage a war to dominate your palate," he said, "and the best beers happen when the two sides become entrenched in defensible positions, protracting the battle into epic proportions." He used the word *quantum* to underscore the "level of seriousness" that he feels ought to be attached to the analysis of beer. Another reason is that, despite his best efforts, he has found it almost impossible to make home-brewed beer consistent in taste and quality—forcing him to conclude that some mysterious, unknown factors in beer production must be operating at the quantum scale.

Other uses are more esoteric. Consider the psychotherapist Lawrence LeShan's defense of mysticism in his book *The Medium, the Mystic, and the Physicist.* Mystics seek the comprehension of a different view of reality, LeShan wrote, before adding, "I use the term 'comprehension' here to indicate an emotional as well as an intellectual understanding of and participation in this view . . . In physics this is called the principle of complementarity. It states that for the fullest understanding of some phenomena we must approach them from two different viewpoints. Each viewpoint by itself tells only half the truth."[26]

The philosopher Žižek, again, has done the most to draw out the philosophical value of the complementarity idea. He compares it to the characteristic line of argumentation of the German philosopher G.W.F. Hegel: "they both reject a position which first posits a gap between the knowing subject and the object-to-be-known, and then deals with the (self-created) problem of how to bridge the gap." The solution—for both Hegel and quantum physics, according to Žižek—turns out to involve incorporating the position and role of the human subject into the relation to what it is seeking to know.[27]

> But in the end, in the end, remember, we have to be able to explain it all to Margrethe!
> —Bohr character, in Michael Frayn's *Copenhagen*

Margrethe Bohr (1890–1984) was born in a small town south of Copenhagen and was intent on becoming a French teacher when she met Niels Bohr in 1910. She married Niels two years later. She was not only Bohr's constant companion but also his intellectual collaborator, a sounding board who helped him revise letters and essays, and to explain ideas to himself. No, it was not possible for Bohr to explain to Margrethe (and therefore himself) what happens in the quantum world in entirely classical terms. We are not sure whether he ever tried. Still, we can imagine what such an explanation would have to look like. To explain it to her would require being resourceful and creative, and appreciating that Margrethe was very smart.

Bohr would have had to begin by saying that, for many historical reasons, today's human beings have inherited a picture of the world, the Newtonian framework, which assumes that things are either waves or particles. Bohr would then have to point out that experimental results have forced us to modify that picture, and why. He would explain that quantum mechanics reveals the existence of another kind of phenomenon that could be either a wave or a particle depending on the experiment we set up to study it, and that all attempts to reduce quantum phenomena to one or the other fail. In just the way that, based on new and unexpected information, we sometimes have to modify our views of close friends and intimates whom we thought we understood well—in ways that we feel ourselves resisting due to our long acquaintance—we now have to do the same about nature. Bohr could then elaborate by saying that this is somewhat like in ordinary perception when the way something appears changes depending on how we observe it— with the difference that the way we observe a quantum phenomenon actually changes what kind of thing the object is, and not just its properties. He could conclude that this is also somewhat like what happens to human beings when certain decisions they make—whom they marry, their career—change them in irreversible ways. Complementarity is not a property of the world, but of the way we have to describe the world in nonmathematical terms.

Assuming that Bohr is sufficiently imaginative—of which we have no doubt—we imagine Margrethe nodding in the end.

Complementarity, Objectivity, and the Double-Slit Experiment

B ohr wrote in his article introducing the notion of complementarity to his colleagues that "an independent reality in the
ordinary physical sense can neither be ascribed to the phenomena
nor to the agencies of observation." This remark made many of his
colleagues go ballistic—and it is still widely misunderstood today.
What can we make of it?[28] Bohr is neither a mystic nor an idealist.
He is not saying that reality does not exist, nor that consciousness
creates reality. He is simply saying that quantum phenomena cannot be separated from the ways that physicists study it. He's being
a hard-headed, empirical scientist.

Let's compare it to the usual assumption of how instruments work.
A Newtonian astronomer confidently uses a telescope assuming that
it improves the ability to see planets and stars, but does not fundamentally transform our vision. What one astronomer sees through a telescope is the same as what another astronomer sees through another
telescope, apart from uncertainties due to skill and ambient conditions. The instrument simply extends our senses. The telescopes can be "dropped out" of our picture of heavenly bodies.
What they show us affects only the epistemology of the heavenly
bodies; what we know of them.

Modern astronomers, to be sure, use different kinds of telescopes to cover different bands of the electromagnetic spectrum
and different amounts of sky acreage. No single instrument fits
all purposes to give us THE heavenly body. The data an astronomer collects depend on the telescope being used; the data are

a function of the instrument and its interaction with the object. Philosophers know that this is an elementary point in perception: how an object—a cup, say—shows itself depends on where we are standing and how we approach it: there is no single stance or "view from nowhere" from which to perceive the entire cup in all of its aspects, and we see it only through ever-changing yet systematically related aspects. Philosophers, like scientists, prefer to use technical terms for what can only loosely be captured in ordinary language, and in philosophical jargon this basic perceptual fact is known as the noetic-noematic correlation: what an object shows of itself to us—the noema—depends on how it's being observed—the noesis. As each changes so does the other. Yet for most purposes our stance can be dropped out, and we can speak of the cup as if it were an independent reality, and as if our different stances affect only epistemology.

The quantum, Bohr pointed out, throws a wrench into this picture, at least in the case of quantum phenomena. The quantum, he wrote in *Nature*, introduces "an essential discontinuity," an intrinsic uncertainty, into atomic processes, "completely foreign to the classical theories." How we access atomic phenomena fundamentally alters how they appear to us. It matters whether we look at position first, or momentum first. It's as if atomic phenomena were abstractions, and if we arrange our instruments to look for particle-like behavior we find it—but if we arrange them to look for wavelike behavior we find *that*. Our instrumental setup—what Bohr called the "agencies of observation"—thus affects not just epistemology but ontology. It would be as if astronomers could look at a heavenly body through a planet-telescope and see a planet, but through a comet-telescope and see comets. You couldn't then take the telescope "out" of the observation, or out of the "evidence" that you are using the instrument to collect. The instrument does more than extend the senses. To make sense of data and synthesize it with other data about the heavenly body in a way that you could communicate to others you would have to describe the entire system of telescope-and-body together, not separately.

As Bohr would put it ten years later, summarizing his Como conclusions, atomic physics has taught us "that the unavoidable interaction between the objects and the measuring instruments sets an absolute limit to the possibility of speaking of a behavior of atomic objects which is independent of the means of observation." There's no way to "get behind" the instruments to see what's "really" there. This produces "an epistemological problem quite new in natural philosophy," which so far has assumed "that it is possible to distinguish sharply between the behavior of objects and the means of observation." This assumption is embedded in language, justified in ordinary experience, and "constitutes the whole basis of classical physics." Its basis, that is, included the assumption that you could speak "from nowhere," observing nature as if it were inside the fishbowl and you were outside. A position of pure spectatorship seemed the very definition of objectivity. Bohr's insight is that quantum mechanics shows this to be false. In the quantum world, where "no result of an experiment . . . can be interpreted as giving information about independent properties of the objects," any result "is inherently connected with a definite situation in the description of which the measuring instruments interacting with the objects also enter essentially."[29] Even scientists have to speak from the standpoint of an experimenter, from the position of someone using equipment and speaking to others. Objectivity thus means what every *experimenter* could in principle see.

In our discussions and in Goldhaber's course on foundations of quantum mechanics, we have come to a new way of expressing Bohr's notion of complementarity. The problem is that both particle and wave phenomena in classical physics provide complete, precise descriptions, in which one may make observations with unlimited accuracy, and use those observations to predict later observation also with accuracy that has no limit in principle. In the quantum mechanics of particles considered one at a time, one can predict the evolution of a wave with unlimited accuracy, but that wave is not observable. The wave is complex, and complex quantities are by definition not real and hence not observable. (So how do we know the waves are there if we cannot observe them? We infer their presence from the patterns left by the detection of many particles. The particles are the "trees" that tell us of the forest.) On the other hand particles *are* observable—for example, a photon striking a given spot on a photographic plate—but their motion is not directly predictable. Instead, the square of the wave function gives probabilities for finding the particle in any given small region of space. Thus the complementarity of particle and wave aspects is simply expressed by saying that quantum mechanics indeed combines the two descriptions—but includes only the predictability of the waves and the observability of the particles.

The most famous, and haunting, demonstration of this is the double-slit experiment, an experiment that was voted the "most beautiful experiment in science," in a poll conducted by Crease in *Physics World* in 2002. In it, electrons are sent, one by one, toward a barrier with two slits on it, and detected on the other side. Even though they cross the barrier singly, the pattern they make when detected shows that they passed through the slits as waves, interfering with each other, and only turn into particles when reaching the detector. This experiment, Richard Feynman declared, exhibits "the only mystery" of quantum mechanics. And one respondent to Crease's poll wrote:

I saw it during an optics course at Edinburgh University. The prof didn't tell us what was going to happen, and the impact was tremendous. I cannot remember the experimental details any more—I just remember the distribution of points that I suddenly saw were arranged in an interference pattern. It is utterly arresting in the way a masterpiece of art or sculpture is arresting. Seeing the two-slit experiment done is like watching a total solar eclipse for the first time: a primitive thrill passes through you and the little hairs on your arms stand up. *Christ*, you think, *this particle-wave thing is really true*, and the foundations of your knowledge shift and sway.[30]

We may express all this in a sentence: John Wheeler summarized Einstein's general relativity (the theory of gravity and particle motion) by saying, "Matter tells space how to curve, and space tells matter how to move." In our case we might say, "Particles (that is, observations of particles) tell waves when, and where, to begin (or end), and waves tell particles where to be."

Chapter Nine

No Dice!

Nooooooooo!!!!!!!!!! This can't beeeeeeee!!!!!!!!!!

That is how some eminent scientists reacted to features of quantum theory as one appeared after the other: discreteness, leaps, randomness, identity, wave-particle schizophrenia, the uncertainty principle, probability waves, and complementarity. Surely such ideas are not *really* true, these scientists felt; what's happening is that we just don't *know* enough about the subatomic world yet. Once we learn more the weirdness will vanish and the Newtonian world will return. Einstein was foremost among the contrarians who came to play an important role in the conceptual development of quantum physics.

We have to pause and talk about this determined resistance, for two reasons. First, the resistance movement plays a role in the development of certain quantum concepts still to come, including Schrödinger's cat and parallel worlds. Second, because this movement, too, has had a cultural impact. But beware! Discussing Einstein's clever objections requires delving into some technical details.

No Way

Many principles of physics have the form: "If you do *this*, what will happen is *that*." Newton's second law of motion, for example, says that the acceleration of a certain mass will be proportional to the force applied to it. Such principles also imply practical impossibilities. You can't do this *unless* you have *that*.

A small number of principles belong to a different category. These say, in effect, "*That* cannot happen." Such principles imply physical impossibilities. You can't do that *at all*.

Notorious examples of such principles include the three laws of thermodynamics. The first says that energy cannot be created or destroyed (energy-wise, "You can't win; at best you can break even"). The second can be stated in several forms, for example, heat cannot be transferred from a colder to a warmer body, or the entropy of a closed system always increases ("You can't break even, except at zero temperature"). The third law finishes it off completely ("You can't get to zero temperature"). Other examples include relativity principles regarding the impossibility of recognizing absolute velocity and the prohibition against faster-than-light travel, and Heisenberg's uncertainty principle.

Such principles do not always represent "new physics," but deductions from other principles. Heisenberg's uncertainty principle is an example, because it is an implication of the "mysterious equation" $pq - qp = Ih/2\pi i$. What is different about the principle is its form. The form is provocative. To say that something is physically impossible tends to make physicists want to rebel, inciting them to seek loopholes.

Fifty years ago, the mathematician and historian of science Edmund Whittaker referred to "postulates of impotence," which assert "the impossibility of achieving something, even though there may be an infinite number of ways of trying to achieve it." Whittaker wrote, "A postulate of impotence is not the direct result of an experiment, or of any finite number of experiments; it does not mention any measurement, or any numerical relation or analytical equation; it is the assertion of a conviction, that all attempts to do a certain thing, however made, are bound to fail."[1]

Postulates of impotence resemble neither experimental facts nor mathematical statements true by definition. But they are fundamental to science. Thermodynamics, Whittaker said, may be regarded as a set of deductions from its postulates of impotence: the conservation of energy and of entropy. It may be possible, he argued, that in the distant future each branch of science will be presented, à la Euclid's *Elements*, as grounded in its appropriate postulate of impotence.

The physics of impossibility goes by several names. "Forget-about-it" physics is one; "no-way" physics is another. No-way physics is important because it attracts contrarians. We are not talking about the endless attempts by frauds and naifs to get around the laws of thermodynamics by creating perpetual-motion machines. Rather, we mean serious physicists who find no-way physics a challenge to devise loopholes. In seeking these loopholes, they end up clarifying the foundations of the field.

One famous example of contrarian physics was James Clerk Maxwell's thought experiment involving a tiny creature who operates a small door in a partition inside a sealed box. By opening and shutting the door, "Maxwell's demon"—as it was later called—lets all the faster-moving molecules into one side of the partition, violating the second law of thermodynamics by getting heat to flow to that side. The discussion of this thought experiment helped to clarify the then-mysterious concepts of thermodynamics.

Contrarian physicists also played a role both in the discovery and in the interpretation of the uncertainty principle. In 1926, Werner Heisenberg was promoting his new matrix mechanics—a purely formal approach to atomic physics—by claiming that physicists had to abandon all hope of observing classical properties such as space and time. The German physicist Pascual Jordan played the contrarian by devising a thought experiment to get around such claims. Suppose, Jordan said, one could freeze a microscope to absolute zero—*then* one could measure the exact position of an electron, say, or the time of a quantum leap. This inspired Heisenberg to think about the interaction between the observing instrument and the observed situation, putting him on the path to the uncertainty principle. Jordan, the contrarian, forced Heisenberg to think operationally rather than philosophically, and to clarify what the physics of the situation was.

Once quantum mechanics appeared, how to connect it with the familiar world that appears to behave Newtonian was one of the great intellectual issues of the early twentieth century. The general strategy worked out by Bohr, Heisenberg, Born, and others, later frequently called the Copenhagen Interpretation, divides the world into two very different domains: one quantum, the other classical. The quantum domain is governed by a field or wave—which is not itself tangible or observable—described by Schrödinger's equation, a recipe that tells the probabilities of certain real states materializing. When this quantum wave encounters something in the classical domain, through a measurement or other interaction, the encounter evaporates or "collapses" the function, eliminating all but the one real state. According to the Copenhagen Interpretation, all we can ultimately know of the world prior to acts of observation or measurement is a set of probabilities.

This interpretation is itself sufficiently strange that it inspired deniers from the beginning. The first, most famous, and intellectually strongest challenger was Albert Einstein. Einstein's objection is often mistakenly cast as "a stubborn old man's nostalgic attachment to classical determinism." But it was far more profound than that; Einstein represented an entire philosophical attitude toward realism, the position that there are things in the world not of our own making, independent of human perception and thought, and that scientific theories are true if they faithfully correspond in some way to these things.[2] By agitating for that realism, Einstein played the contrarian role—with Bohr as his principal adversary—and devised clever ways of trying to simultaneously determine the position and momentum of a particle assuming that these properties existed apart from how we measure them. While all his attempts failed, the discussion it provoked did much to help physicists understand the nature and implications of quantum mechanics.

Einstein the Contrarian

Einstein expressed his doubts early and often. In his quantum theory papers of 1916, he wrote that it is a weakness of his theory "that it leaves time and direction of elementary processes to chance." However, he expressed "full confidence" in his arguments. This confidence was halfhearted at best. Eight years later, reacting to the ill-fated Bohr-Kramers-Slater (BKS) attempt to keep wave theory by abandoning the conservation of energy, Einstein wrote to Born that "I should not want to be forced into abandoning strict causality without defending it more strongly than I have so far. I find the idea quite intolerable that an electron exposed to radiation should choose of its own free will, not only its moment to jump off, but also its direction. In that case, I would rather be a cobbler, or even an employee in a gaming-house, than a physicist."[3] Two years later, after Heisenberg and Schrödinger produced their two maps of the quantum domain, Einstein wrote to Born that he was impressed, but added that "an inner voice tells me that it is not yet the real thing," and that the theory doesn't bring us closer "to the secret of the Old One." Einstein then made a now-famous and often-repeated comment: "I am at all events convinced that HE does not play dice," often quoted with the additional phrase (which Einstein did not use) "with the universe." Two years later, he wrote Sommerfeld that the Heisenberg-Dirac theories "don't have the smell of reality."[4]

The next year, 1927, was both the bicentennial of Newton's death and the year the uncertainty principle appeared. This made it undeniable that quantum mechanics had given birth to an epistemological crisis, and brought about the first serious explorations of the philosophical implications of the new order. Einstein, however, was still hoping for a restoration of the old. In an article in *Nature*, published in March, he wrote,

> It is only in the quantum theory that Newton's differential method becomes inadequate, and indeed strict causality fails us. But the last word has not yet been said. May the spirit of Newton's method give us the power to restore unison between physical reality and the profoundest characteristic of Newton's teaching—strict causality.[5]

In the next few years, Einstein would try with all his might to restore that unison.

Back to Reality

Einstein's struggle unfolded in several phases. Throughout, he was an "honest" objector to quantum mechanics in that he valued its structure and appreciated the experimental facts on which it was based. He was not a "denier," not averting his eyes from scientific evidence. He realized that to prevail he would have to find a more complete, fully causal way of looking at things, and phrased his arguments accordingly. His thinking in each phase crystallized around a thought experiment. While Einstein did not invent thought experiments—Galileo, for instance, was also an imaginative thought experimenter—he was a champion at them. His earliest and most famous came to him as a teenager: he imagined being a "light surfer," riding the crest of a light wave. That would mean viewing the light wave as standing still, but Maxwell's equations guarantee that light in a vacuum always travels at the speed c regardless of how it's seen. Einstein fashioned his special theory of relativity to resolve this contradiction. His thought experiments attacking quantum mechanics were no less imaginative.

ENSEMBLE INTERPRETATION

The first phase began in 1927, after Heisenberg enunciated the uncertainty principle. That May, Einstein presented a paper entitled "Does Schrödinger's Wave Mechanics Determine the Motion of a System Completely or Only in the Sense of Statistics?" to the Prussian Academy of Sciences. In the surviving manuscript, he maintained that "the assignment of completely determined motions to solutions of Schrödinger's differential equation" is indeed possible just as it is in classical mechanics.[6] Einstein appears to have intended to present something like this idea in Brussels at the upcoming, fifth, Solvay conference to take place in October 1927. But he withdrew the paper from publication—possibly because of criticisms by Heisenberg—and announced that he would not present anything in Brussels.

Einstein did not attend the Como conference that September, at which Bohr outlined his views on complementarity, but did attend the Fifth (1927) Solvay Conference a month later. This has been called "perhaps the most important meeting in the history of quantum theory," at which "a range of sharply conflicting views were presented and extensively discussed, including de Broglie's pilot-wave theory, Born and Heisenberg's quantum mechanics, and Schrödinger's wave mechanics."[7] Louis de Broglie sympathized with Einstein, and his formal presentation offered an alternative interpretation of quantum mechanics that would restore determinism by postulating that Schrödinger's equation amounted to a guiding or "pilot wave" for particles whose exact trajectories might be determinable with further information.[8] Einstein gave no formal presentation, but expressed his opposition to quantum mechanics during the general discussion. "Despite being conscious of the fact that I have not entered deeply enough into the essence of quantum mechanics," he began modestly, "nevertheless I want to present here some general remarks."[9] He drew a diagram of a screen punctured by a small hole, on the back of which was a hemisphere of photographic film. Electrons strike the screen, some of which pass through the small hole and are diffracted, spreading out uniformly in all directions, the wave function for which is ψ.

We can consider what is happening in two ways, Einstein continues. In quantum mechanics as envisioned by Bohr, Heisenberg, and most others, ψ is a complete description of the individual processes, what's happening to an electron, say. The expression $|\psi|^2$ represents the probability that the electron is found at a given time at a given point on the screen. It's possible, Einstein says, that the electron could be found at several points on the screen; but if it turns up at one point, that prevents the electron from turning up at another. This process "assumes an entirely peculiar mechanism of action at a distance," for it "prevents the wave continuously distributed in space from producing an action in *two* places on the screen."[10] An action that occurs over *here* makes it impossible for an action to occur over *there*.

But we can remove that objectionable mechanism, Einstein continues. Suppose that ψ does not represent what's happening to a single particle but to an *ensemble* or cloud of systems or particles. The theory is then incomplete in a similar way that thermodynamics is incomplete, for it addresses the average behavior of a large number of processes rather than a single one. A more detailed theory might be able to describe specific trajectories of particles, rather than just the aggregate behavior.

Bohr, perhaps being charitable, professed confusion at Einstein's remarks: "I don't understand what precisely is the point which Einstein wants to [make]."[11] His confusion was perhaps understandable, for instead of proposing a pilot-theory–like view Einstein was doing something different: pointing out that the price of Bohr's position—of saying that quantum mechanics is complete—is admitting action at a distance, or what would later be called "nonseparability" or "entanglement"—that is, that an action *here* can instantaneously affect an action *there*. Another way to say this is that the action is not local, but can occur simultaneously even in widely separated places.

Bohr would soon learn Einstein's position at great length. This exchange at the Solvay conference in 1927 inaugurated a lengthy, now-famous struggle between Einstein and Bohr, carried on initially in private discussions and soon more publicly, and the two remained sparring partners for the rest of their lives.

Bohr brooded about what might be called Einstein's ensemble interpretation. It is often called a statistical interpretation, but confusingly this is also the name for Born's completely different interpretation of quantum mechanics, so we'll call it the ensemble interpretation. Also, Einstein's point was not really about statistics in Bohr's account but about nonseparability, that in Bohr's theory two spatially separated objects, or properties of objects, could be treated as connected. Einstein felt sure that this was an undesirable feature of the theory and would be eliminated once it was complete. Over the next few years, Bohr offered several counters to Einstein's ideas.

Photon Box Experiment

The second phase of Einstein's struggle occurred at the Sixth Solvay Conference in 1930. The conference was nominally about magnetism, but that did not deter the participants from discussing what they pleased. Einstein would come to breakfast with some objection to quantum mechanics; Bohr would think about it, and then return with an answer by dinner.

Einstein presented a thought experiment that looked as though it could trick a photon into showing both its exact energy and the time at which that energy was measured. This was impossible according to the uncertainty principle, which would apply also to the relation between energy and the time at which it was measured. The experiment involved creating a box that could be very carefully weighed, and that could allow a photon to escape through a shutter left open for a very brief time. The exact time that the photon escaped could be pinned down by a clock controlling the shutter, while the photon's exact energy could be precisely determined by weighing the box before and after. This device, Einstein thought, would enable an experimenter to measure both the photon's energy and the time of its escape, contradicting the energy-time uncertainty relation.[12] As one physicist present later wrote:

> It was quite a shock for Bohr . . . he did not see the solution at once. During the whole evening he was extremely unhappy, going from one to the other and trying to persuade them that it couldn't be true, that it would be the end of physics if Einstein were right; but he couldn't produce any refutation. I shall never forget the vision of the two antagonists leaving the club: Einstein a tall majestic figure, walking quietly, with a somewhat ironical smile, and Bohr trotting near him, very excited. . . . The next morning came Bohr's triumph.[13]

Bohr interpreted Einstein's thought experiment as an attempt to defeat the uncertainty principle and trick nature into revealing both the energy of the photon and the time at which it was measured. His refutation then involved pointing out that weighing the box requires use of a spring to balance against gravity, and accurate determination of the change in vertical position of the box, which means a corresponding uncertainty in the vertical momentum of the box. That in turn would lead to a shift in clock rate (according to Einstein's famous "gravitational red shift," a fundamental feature of his general theory of relativity). When all these effects are put together one gets the required uncertainty product for photon energy and photon emission time.

EPR PAPER

Einstein still did not give up. In February 1931, he submitted a short piece entitled "Knowledge of Past and Future in Quantum Mechanics," coauthored by Richard Tolman and Boris Podolsky, to the editor of *Physical Review*. It's another blast at the uncertainty principle using another thought experiment, pointing out that keeping it requires junking the presumably essential qualities of locality and separability. "It is well known that the principles of quantum mechanics limit the possibilities of exact prediction as to the future path of a particle. It has sometimes been supposed, nevertheless, that the quantum mechanics would permit an exact description of the past path of a particle." But a "simple ideal experiment," the authors continued, would show "that the possibility of describing the past path of one particle would lead to predictions as to the future behavior of a second particle of a kind not allowed in the quantum mechanics." The article does not say that quantum mechanics is wrong, only that keeping the uncertainty principle comes with what the authors felt was unacceptable baggage—it means that you cannot have precision in the past, either.[14]

Einstein was more blunt in public after the paper was submitted. I know what I am about to say will be interpreted as "a sign of senility," he told a *New York Times* reporter, but causality will return, and scientists will one day recognize it again as "an inevitable part of the nature of things." Einstein said, "The mere 'probable' correctness of the present mathematical equations is only an indication of the fact that these new conceptions are still missing." I feel, he continued, that the "essentially statistical character" of quantum mechanics "will eventually disappear because it leads to unnatural descriptions, since it does not describe nature but merely expectations from nature, while the aim of science is describing the things themselves, not merely the probability of their happening."[15]

Ah, the things themselves! Einstein and Bohr were arguing philosophically about the nature of "things." Were things always localized in space and time? Or were there "things" that could not be localized? If so, what kind of a thing could that be?

The next day, the *Times* editorialized about Einstein's "conservatism." His objection to quantum theory is based solely on "a natural repugnance," the editors wrote. Scientists aren't usually motivated by repugnance, and Einstein understandably fears being labeled senile. Still, that reaction to quantum mechanics is "human," the *Times* admitted. "Something is clearly wrong."[16]

In 1932, in a book called *Mathematical Foundations of Quantum Mechanics*, the mathematician John von Neumann offered a proof of a theorem that seemed to eliminate any possibility of a theory of quantum mechanics based on hidden variables. "Hidden variables" here referred to quantities not yet known to us whose discovery might permit exact predictions, rather than just deductions of probabilities, for outcomes of experiments in which quantum mechanics plays a central role. Quantum mechanics, it seemed, was not a temporary theory awaiting more work but stated all that we could know.

Einstein and a few other contrarians saw this proof as a challenge. In 1933, he left Germany for the United States after the Nazis took power in his home country. He settled in Princeton, at the Institute for Advanced Study, and embarked with renewed passion on showing that quantum mechanics was incomplete, together with two other Princeton physicists—Podolsky and Nathan Rosen, who had become Einstein's assistant. Reporters periodically interviewed Einstein on his views on quantum mechanics, knowing that they could count on getting a good quote. In 1934, for instance, he told an assembly of reporters that in his eyes, the extension of the uncertainty principle to human life on a large scale was "a matter of taste." A *Times* report on the gathering continued, "For his part he preferred to believe that everything in nature and in life was governed by cause and effect," adding that he hoped future science would bear him out.[17]

Einstein soon embarked on a paper that he hoped would start to get things back on track. On May 4, 1935, the *New York Times* ran an article entitled "Einstein Attacks Quantum Theory; Scientist and Two Colleagues Find It Is Not 'Complete' Even Though 'Correct.'" Like Einstein's previous attempts, this one was motivated by the view that there was a real state of a system, something that existed independently of whether it was measured or observed. Because quantum mechanics could not describe such a real state, it had to be incomplete.

Quantum mechanics, Einstein was implying, was something like a map. A map, too, is correct but incomplete. It doesn't give you a landscape, but rather an abstract picture of a terrain that leaves out much information about underlying features such as altitude or soil character. If you could look at the underlying features you could get a much better picture of the terrain without contradicting the map. Just so, Einstein seems to have thought, if you could look at the underlying features of the subatomic world, you would get a picture much more in accord with Newtonian intuitions without contradicting quantum mechanics.

The result was "Can Quantum-Mechanical Description of Physical Reality Be Considered Complete?" which Einstein coauthored with Podolsky and Rosen. Universally known as the EPR paper, it was published in May 1935.[18] It has become one of the most famous scientific papers of all time. In the very first sentence it invokes reality:

Any serious consideration of a physical theory must take into account the distinction between the objective reality, which is independent of any theory, and the physical concepts with which the theory operates. These concepts are intended to correspond with the objective reality, and by means of these concepts we picture this reality to ourselves.

Theories picture a reality that exists entirely separately from the picturing. A satisfactory theory, the authors continue, answers two questions: (1) "Is the theory correct?" and (2) "Is the description given by the theory complete?" The authors don't bother with (1); they, and everyone else, knew that every prediction of quantum mechanics had been experimentally confirmed and that nothing in this respect was likely to change. Regarding (2), they defined what they called the "condition of completeness" in the following way: "every element of the physical reality must have a counterpart in the physical theory." In a complete picture of a house, say, each component of the picture must be of something in the house itself. The authors provide a sufficient, though not necessary, definition of reality as follows:

If, without in any way disturbing a system, we can predict with certainty (i.e., with probability equal to unity) the value of a physical quantity, then there exists an element of physical reality corresponding to this physical quantity.

If we can tell from a complete picture of a house, say, where a particular part of the house is that corresponds to a component of this picture, then that is a "real" part of the house. That is reasonable. Now, in quantum mechanics, the wave function ψ completely characterizes the state of a particle, which is a function of the variables used to describe the particle. The paper then goes on to show that when the momentum of a particle is known, its position has no physical reality. This simply restates Heisenberg's uncertainty principle.

But, they continued, we can set up experimental situations where we can know more than that! Suppose we produce two electrons whose momenta and positions are linked, which we can do by de-excitation of some atom at rest. We can measure the momentum of the first and know the momentum of the second, but we could also measure the position of the second and then know the position of the first. Thus the two supposedly incompatible quantities of each particle could both be known, even though from the uncertainty principle both could not be "real" at the same time in the sense that the EPR paper had defined. In creating this argument, they were seeking to restore the commonsense view that both particles had positions and momenta from the moment of their creation, independently of each other. In fact, our very definition of reality, Einstein and his coauthors held, *depends* on there being elements in such separated systems whose existence is independent of observation and measurement. This contradicts the story told by Bohr, Heisenberg, and company! The answer to the rhetorical question of the title was therefore No.

"We are thus forced to conclude," the authors wrote, "that the quantum-mechanical description of physical reality given by wave functions is not complete." Position and momentum cannot depend on measurement. The wave function therefore "does not provide a complete description of the physical reality." Is there an alternative that does? We think so, Einstein and company said.

When the *Physical Review* issue containing the EPR paper arrived in Copenhagen, Bohr reacted badly, finding it to be an unwelcome bolt from the blue. Regarding it as a threat to the foundations of quantum mechanics, Bohr immediately put down his other research and worked for six intense weeks on a reply. This inspired Einstein to continue work on the idea that quantum mechanical wave functions only describe an ensemble of systems, and do not yet describe the underlying features of the subatomic world. Einstein kept developing this viewpoint until the end of his life, in favor of causality and opposition to nonlocality, which in 1948 he called "spooky action at a distance" (in the original German, *spukhafte Fernwirkung*—a mouthful!). In 1949 Einstein wrote, "I am, in fact, fully convinced that the essentially statistical character of contemporary quantum theory is solely to be ascribed to the fact that this [theory] operates with an incomplete description of physical systems."[19] Bohr replied along the lines he had for the previous two decades: "[I]n quantum mechanics, we are not dealing with an arbitrary renunciation of a more detailed analysis of atomic phenomena, but with a recognition that such an analysis is *in principle* excluded." While Einstein continued to argue that our notion of reality entails that quantum mechanics is incomplete, Bohr continued to argue that Einstein's notion of reality was wrong.

Night Thoughts

Stories and plays involving characters who are undone by resisting change and sticking to tradition go back to ancient times, and appear today everywhere from the Nigerian novelist Chinua Achebe's book *Things Fall Apart* to the Broadway musical *Fiddler on the Roof*. We have mentioned several scientists—including Millikan—who clung to Newtonian mechanics for a while and converted to quantum theory against their better judgment. We've even seen one physicist, Arthur G. Webster of Clark University, whose suicide may have been due partly to his despair over the intellectual and institutional changes in physics wrought by the quantum.

At least one novel has this very issue at its core: *Night Thoughts of a Classical Physicist*, by Russell McCormmach, a historian of physics. The protagonist, Viktor Jacob, is a classical physicist at a minor institute in a minor German university. Jacob, born in the mid-nineteenth century, had been "drilled in the classics, in the careful thought of the languages and literatures of antiquity" in his youth. He was attracted to physics for the way "it connected physical facts through a small number of exact, fundamental concepts and laws." Now, everything is crumbling—the physicists' picture of light, and with it, as if in resonance, everything else. Jacob comes up with an idea, which he calls the "world-ether," that he hopes will restore intelligibility to physics. He visits Planck to pursue the idea; the great man shows him "guarded sympathy," but in the end seeks to dash Jacob's hopes. Jacob feels that the crumbling of his personal life is tied to that of German academic life and somehow even to the advent of the quantum. At the end, on the eve of the German defeat in the First World War, Jacob evidently kills himself.[20]

The Sacrifice

Heisenberg once wrote: "Almost every progress in science has been paid for by a sacrifice, for almost every new intellectual achievement previous positions and conceptions had to be given up. Thus, in a way, the increase of knowledge and insight diminishes continually the scientist's claim on 'understanding' nature." Heisenberg is overstating: surely the advance of science involves developing more subtle and complex concepts that encompass the simpler existing ones. But these more subtle and complex concepts are often produced by those who are dissatisfied by the prospect of having to make the kind of sacrifice Heisenberg mentions.

Dissatisfaction, indeed, is a powerful driving force in science, and it can arise in many ways. Sometimes it springs from a scientist's sense that a confusing heap of experimental data can be better organized. At other times it arises from the feeling that a theory is too complicated and can be simplified, or that its parts are not fitting together properly. Still other dissatisfactions arise from mismatches between a theory's predictions and experimental results.

No-way physics produces a special kind of dissatisfaction, involving the collision of science with our hopes and dreams—of limitless energy, of superluminal travel, of pinning things to specific places at specific times. Human beings seem hardwired to have such hopes, and hardwired to balk at the science that dashes them. Small wonder that no-way physics leaves them dissatisfied. But science wins in the end.

John Bell and His Theorem

The dispute between Einstein and Bohr was still unresolved when Einstein died in 1955. Nevertheless, the struggle between these two forceful, imaginative, and insightful individuals did much to clarify what quantum mechanics truly meant. The struggle was soon picked up by a new generation. One of its principal members was John Bell (1928–1990), who started out as a quantum contrarian but soon changed his mind, and to his puzzlement and consternation ended up a quantum guru among a certain coterie of nonphysicists.

One of us—Crease—spoke with Bell about his unusual career shortly before Bell's death in 1990. Before the conversation, Bell warned that he was an "impatient, irascible sort" who brooked no nonsense. In person he proved patient and soft-spoken, punctuated by moments of intense passion.

John Bell

One day in 1960, Bell unexpectedly found himself sharing an elevator with Niels Bohr at CERN, the then-new international scientific laboratory outside Geneva, after an opening celebration. The young Irishman, barely in his thirties, began summoning the courage to tell the eminent seventy-five-year-old founder of quantum mechanics, "I think your Copenhagen Interpretation is lousy!" In the brief time the two were alone, Bell lost his nerve

John S. Bell (1928–1990).

for one of the few times in his life. "If the lift had gotten stuck between floors," Bell recalled, "it would have made my day!"[21]

Bell said he was born in Belfast in Northern Ireland in 1928, the child of two relatively poor Protestants. He graduated from high school in 1944, in the middle of the war, but was not conscripted. In Northern Ireland about 60 percent of the population was British and Protestant; 40 percent was Irish Catholic and "only reluctantly British." Conscription, Bell explained, would not have worked because "They couldn't say, 'We will just conscript the Protestants!'" In 1945 he entered Queens University of Belfast, where, for a time, philosophy diverted his scientific interests. But he got frustrated, because "each generation of philosophers seemed to simply overturn the generation before." Physics was not far from philosophy, he decided, and at least its knowledge accumulated. Bell returned to physics, but continued to enjoy throwing darts at philosophers.

Quantum mechanics and the uncertainty principle troubled Bell the moment he learned of them at Queens. He was enraged by the muddled explanations he found in textbooks, and by his professors' inability to clarify matters. When Bell asked his teachers, "Is the wave function real or just a book-keeping scheme?" and they could not answer to his satisfaction, he provoked them, resulting in some unpleasant exchanges. When he accused one of dishonesty, the teacher—equally heated—angrily snapped back that Bell was "going too far."

Bell said he learned quantum mechanics not the way you learn to play chess—by learning rules—but the way you learn to ride a bicycle, "without knowing what you're doing." Ignoring the meaning of what he was doing left him uneasy, but still he was smart enough to know that he had to apply himself to behave like a conventional physicist and solve conventional physics problems if he wanted to have a physics career. Bell resigned himself to practicing his ruminations about the meaning of quantum mechanics as "a hobby."

But Bell pursued his hobby seriously, even after receiving his PhD from Queens in 1949. He read the EPR paper carefully, but was deeply dissatisfied by Bohr's response: "huffing and puffing," he called it, "not a reply." He read von Neumann's "no hidden variables" proof equally carefully, but was outraged. "The literature is full of respectful references to this proof," Bell said, "to the 'brilliant proof of von Neumann.' And I do not believe it can have been read at that time by more than two or three people."

Why not?

"The physicists don't want to be bothered with the idea that maybe quantum theory is only provisional. A horn of plenty had been spilled before them, and every physicist could find something to apply quantum mechanics to. They were pleased to think that this great mathematician had shown it was so. Yet the von Neumann proof, if you actually come to grips with it, falls apart in your hands!"

Bell grew visibly annoyed.

"There is *nothing* to it. It's not just flawed, it's *silly*. If you look at the assumptions made, it does not hold up for a moment. It's the work of a mathematician, and he makes assumptions that have a mathematical symmetry to them. When you translate them into terms of physical disposition, they're nonsense. You may quote me on that: The proof of von Neumann is not merely false but *foolish*!" (So quoted.)

Bell has been accused of exaggerating the flaws in von Neumann's "no hidden variables" proof, which turns out to exclude only certain kinds of hidden variable theories.[22] What's important, however, is that von Neumann's proof had aroused and focused his contrarian instincts. These instincts were further encouraged in 1952, when he read a paper in the *Physical Review*, "A Suggested Interpretation of the Quantum Theory in Terms of 'Hidden Variables,'" in which the author, David Bohm, extended certain proposals that de Broglie had made at the 1927 Solvay conference.[23] "That was a big thing for me," Bell said, "It told me that von Neumann was wrong, because Bohm had done what von Neumann had said to be impossible." But Bohm paid a price for introducing hidden variables; he had to introduce nonlocality into the theory. "That means that what you do here"—Bell points to the desk in front of him—"has immediate consequences in remote places"— pointing out the window to Mt. Blanc. "And that was *extremely* odd." It is what Einstein called "spooky action at a distance." Bell began to brood about whether you could create a hidden variable theory without spooky action at a distance.

In 1960, Bell joined CERN, where he worked on theoretical aspects of the physics experiments being conducted there. Early in 1963, he heard a talk by a physicist visiting CERN who claimed to have improved von Neumann's proof. Another red flag, waved in front of the voluble physicist. "I was very indignant," Bell said. Later that year, he was invited to the Stanford Linear Accelerator Facility (SLAC), a laboratory near San Francisco. He arrived the day after President Kennedy's assassination; "it was a very odd experience to find everybody crushed." At SLAC, with a little extra time on his hands, he decided to pursue his hobby, and he wrote two highly influential papers that, he said, made him "notorious."

The first, "On the Problem of Hidden Variables in Quantum Mechanics," exposed the flaws in von Neumann's proof.[24] It was worth doing, he wrote, because even three decades later, one could not find "in the literature any adequate analysis of what went wrong." But in laying out the argument, Bell found it hard to avoid the conclusion that hidden-variable theories had to be nonlocal, and ended the paper with that suggestion.

Bell set out to settle the issue for good in his second paper, "On the Einstein-Podolsky-Rosen Paradox."[25] He said, "I tried to imagine what hidden variables there might be that would avoid the nonlocality of Bohm . . . And I found that I couldn't do it." A remarkable thing then happened. Instead of looking to disprove the impossibility of a hidden variable theory that would preserve locality, he began to look for its opposite. "I made a phase transition in my mind. I started looking for the proof of impossibility. And I found it."

Bell chuckled as he recalled the episode. "What I wanted was a clear argument, rather than to justify any particular conception of the world. I wanted to sort out the logic of such questions . . . I am often more concerned with the conduct of the debate and with the logic than the actual truth."

This is Bell's most influential paper. In any theory of quantum mechanics, it showed, "the setting of one measuring device can influence the reading of another instrument, however remote"— and do so instantaneously, that is, faster than the speed of light. Within a few years, the theory was developed and extended by others, who pointed out ways that Bell's theorem could be experimentally tested.[26]

Roughly—there are plenty of good and fairly easy to follow explanations, but it takes more space than we have here[27]—Bell's theorem begins with the Einstein-Podolsky-Rosen thought experiment involving two particles such as electrons emitted in opposite directions. But instead of considering position and momentum, it considers another property of such particles, called spin. Instead of considering exact correlations, it considers intermediate settings. When a property of one of them is actualized by being measured, the properties of the other will be found to be correlated—linked— when it is measured, no matter how far apart the two particles have become. The correlation is in the form of an inequality that has one form for classical physics and another for quantum physics. In the quantum case, it looks like the two particles have communicated instantaneously—either that, or that they have formed one large, extended object such that when you tweak one end the other end "feels" it.

"The theorem certainly implies," Bell said, "that Einstein's concept of space and time, neatly divided up into separate regions by light velocity, is not tenable." Within a few years, Bell found himself a celebrity among a strange coterie of people with whom he had never dreamed he would associate: those who thought that physics taught the same lessons as Eastern mysticism. Bell chuckled, responding,

> I am essentially an agnostic about religious and spiritual matters. When people give answers to these questions, I think it's wishful thinking. I don't feel hostile toward these people, but I just don't share their enthusiasm at finding answers to questions that seem to me unanswerable. I admit that there are questions that science cannot answer—that science cannot even ask. But I myself don't have answers to these questions. When I hear people saying we've finally answered it, and the answer is Buddhism or Taoism or something else, I just have to say that when I look at those things I don't find the answer. Even so, if other people find the answers there, I'm not going to campaign against them. That's their business. They are doing no great harm. There are ideologies that are much more vicious than Buddhism.

Nothing irascible about that.

Chapter Ten

Schrödinger's Cat

"Can you explain to me about Schrödinger's cat?"
"Come on," she says, and reaches out for my coat and pulls me down the sidewalk. I'm walking beside her, not saying anything, and she's mumbling, "God God God God God God God," and I say, "What's wrong?" and she says, "You, You, Grayson. You're what's wrong," and I say, "What?" and she says, "You know," and I say, "No I don't," and she—still walking and not looking at me—says, "There are probably some girls who don't want guys to show up at their house randomly on a Tuesday night with questions about Edwin Schrödinger. I am sure such girls exist. But they don't live at my house."

—*Will Grayson, Will Grayson*

Will and Jane, the two geeky teenagers in *Will Grayson, Will Grayson* that we met in Chapter 1, talk about Schrödinger's cat with ease. To catch up with them, we need a brief review, some technical detail, and to delve into the original paper in which Schrödinger introduced the idea.

Quantum mechanics, as we have seen, describes the world as the product of two ingredients. The first is an information function, the ψ-function described by Schrödinger's equation, a classical wave that expands outward and overlays, or "superposes," many possibilities. The second ingredient is something that befalls this function, causing it to disappear and one of its possibilities to appear.

How does this strange picture connect with the familiar everyday world? Answering this question—"interpreting" quantum mechanics—is, we have said, one of humanity's greatest intellectual challenges. The pioneers of quantum mechanics were perplexed, and when expressing themselves nontechnically were all but helpless, and usually resorted to metaphors or handwaving. Eddington said that the electron "crystallizes out of Schrödinger's mist like a genie emerging from his bottle," Bridgman wrote that the physicist inhabits "a world from which the bottom has dropped clean out," while Bragg produced the image of a wave that disperses over thousands of miles but when it strikes a ship it can mysteriously summon its total energy, now spread out over a huge area, to blast a piece of timber clear out of the hull. Baffling!

More technically, physicists at the time had many different ways of interpreting what was happening.[1] The most widely known interpretation—Einstein, Schrödinger, and other skeptics called it "dogmatic quantum mechanics" and much later the "Copenhagen Interpretation"—was crafted mainly by Bohr and Heisenberg in the late 1920s and early 1930s. It sees the world as divided into two separate domains: quantum and classical. The quantum domain is governed by the unobservable and intangible ψ-field described by Schrödinger's equation, a recipe that tells the probabilities of certain real states materializing. When this quantum wave encounters something in the classical domain, through measurement or other interactions, the encounter evaporates or "collapses" the function, one hitherto only probable state now becomes "real," and the other possibilities are eliminated. According to this interpretation, all we can ultimately know of the world prior to acts of observation or measurement is a set of probabilities as to which states might be realized.

This idea, we saw in the previous chapter, was sufficiently weird to spark opposition from the beginning. Einstein was the chief combatant, and his largest salvo was fired in the EPR paper, "Can Quantum-Mechanical Description of Physical Reality Be Considered Complete?" published in May 1935.

In 1935, Schrödinger was an exile in England. Born in Austria, he became a professor in Berlin in 1927, succeeding Max Planck at the Friedrich Wilhelm University. In 1933, however, he had been angered and then terrified by a series of measures adopted by the new Nazi government, including racial laws and book burnings, and he left, eventually winding up in Oxford. When the *Physical Review* issue with the EPR paper arrived in Oxford, he was thrilled. On June 7, 1935, he wrote to Einstein expressing his delight: "you have evidently caught dogmatic q.m. by the coat-tails," using their derogatory term for the Bohr-Heisenberg view denying the reality of certain properties such as position and momentum apart from measurement situations.[2] We are now en route to the cat idea. Einstein replied enthusiastically to Schrödinger, and elaborated his ideas: Physics describes reality, but not all descriptions are complete. Imagine two boxes with lids you can open to peer inside, and there's a ball in one. Before you "make an observation" by looking inside, how do I describe the situation? We say, correctly, that the probability that the ball is in the first box is ½, or 50 percent. Is that a complete description? Of course not, Einstein says. It characterizes only our *knowledge* of the situation, not reality itself. Really, the ball is in the first box or it isn't. Yet according to "dogmatic q.m.," it can in principle be a complete description to say the chance of its being in that box is 50 percent. Dogmatic q.m. seems to be saying that the ball is not in one or the other box, but first exists in a box only when you peer inside. That's ridiculous, Einstein continues. It violates the principle of "separation," according to which what's in the second box is independent of what's in the first. This idea is so essential to our notion of what we accept as real that he calls it the principle of "local reality." In the Bohr-Heisenberg dogmatic q.m. account, Einstein wrote incredulously, "The state *before* the box is opened is completely described by the number ½."[3]

Two months later, Einstein sent Schrödinger another analogy. Suppose a pile of gunpowder has a probability of exploding in a year, he mused. So therefore its ψ-function is a superposition of exploded and unexploded gunpowder? Nonsense! Quantum mechanics with its ψ-function, Einstein declared, is incomplete and inadequate. "There is no interpretation by which such a function can be considered to be an adequate description of reality."[4]

Meanwhile, the EPR paper had reached Pauli in Zürich. Outraged, Pauli exchanged heated letters with Schrödinger. Einstein and company, Pauli said, were backsliding, in danger of envisioning the ψ-function as something like statistical gas theory—probabilistic, yes, but because it deals with a huge number of components. Quantum mechanics, Pauli insisted, decisively ruled out the closet determinism and realism behind the EPR paper.

The EPR paper, followed by the correspondence with his intellectual ally Einstein and opponent Pauli, inspired Schrödinger to set down an informal account of his own views. He called it "general confession," and published it in October 1935 using the title "The Present Situation in Quantum Mechanics."[5] This paper is now famous as the first appearance of Schrödinger's cat, which today regularly appears in popular discussions of the weird realities of quantum mechanics.

"Schrödinger was doing a thought experiment. Okay, so, this paper had just come out arguing that if, like, an electron might be in any one of four different places, it is sort of in all four places at the same time until the moment someone determines which of the four places it's in. Does that make sense?"

"No," I say. She's wearing little white socks, and I can see her ankle when she kicks up her feet to keep the swing swinging.

"Right, it totally doesn't make sense. It's mind-bendingly weird. So Schrödinger tries to point this out. He says: put a cat inside a sealed box with a little bit of radioactive stuff that might or might not—depending on the location of its subatomic particles—cause a radiation detector to trip a hammer that releases poison into the box and kills the cat. Got it?"

"I think so," I say.

"So, according to the theory that electrons are in all-possible-positions until they are measured, the cat is both alive and dead until we open the box and find out if it is alive or dead. He was not endorsing cat-killing or anything. He was just saying that it seemed a little improbable that a cat could be simultaneously alive and dead."

—Will Grayson, Will Grayson

Jane is giving an explanation that is easy for a fellow geek to follow. Schrödinger's original paper is not much more difficult for a determined reader to understand. He begins by saying that the classical world gave us the idea that nature can be exactly described. Sure, experimental data may not allow this description to be carried out in complete detail in practice, but such data do let us model phenomena that we can compare to reality and fix if necessary so our models fit the phenomena better. While some models are static, others are dynamic and describe phenomena as passing in time from one *state* to another in time. By refining our models, thought adapts to experience.

These states are specified by a number of what Schrödinger calls *determining parts* or variables. The determining parts of a (static) triangle, for instance, include sets of its angles and sides: all properties of a triangle can be completely specified by one angle and two sides, or two sides and an angle, or three sides. A small set of these determining parts uniquely determines all the others in a state, but different sets can be used to determine all the other parts.

Yet this is impossible in quantum mechanics. Not all determining parts or variables can be codetermined. The obstacle is not any practical limitation but Heisenberg's uncertainty principle, or the fact that when you measure some variables, others become unmeasurable. The classical notion of a state is now meaningless. Schrödinger now asked, What about those other variables? *"Have they then no reality, perhaps (pardon the expression) a blurred* reality; or are all of them always real and is it merely . . . that simultaneous *knowledge* of them is ruled out?"

In philosophical language, Schrödinger was asking whether the probabilities affect the ontology of the variables—whether the quantities to which they refer *exist* or not—or merely their epistemology, that is, our ability to *know* what they are. Attentive readers of our previous chapters already know the quantum mechanical answer: the former.

In thermodynamics, he continued, probabilities affect only epistemology. Scientists model systems containing billions of billions of molecules by treating them, conveniently yet implausibly, as if they involve single states arbitrarily chosen from ensembles of many possible states. In thermodynamics you don't care how a system behaves exactly—indeed, you aren't even interested in the exact description of each and every atom—only how it behaves for the most part. But "this won't work" in the quantum domain, because some variables must remain indeterminate or blurred when others are known to be exact. Perhaps if we knew more about the underlying situation, Schrödinger said, we would find it more complex than we thought, and causality might reappear. Still, as long as the ψ-function is confined to the subatomic domain the indeterminacy is harmless. "Inside the nucleus, blurring doesn't bother us." However, Schrödinger continues, "serious misgivings arise if one notices that the uncertainty affects macroscopically tangible and visible things, for which the term 'blurring' seems simply wrong." Schrödinger then conjured his famous image, cleverly extending the indeterminacy of the micro domain into the macro world more straightforwardly than Einstein's examples:

> One can even set up quite ridiculous cases. A cat is penned up in a steel chamber, along with a Geiger counter, which must be secured against direct interference by the cat. The Geiger counter contains a tiny bit of radioactive substance, so small,

that *perhaps* in the course of the hour one of the atoms decays, but also, with equal probability, perhaps none. If an atom does decay, the counter tube discharges and—through a relay—releases a hammer that shatters a small flask of hydrocyanic acid. But if no atom decays after an hour, the cat still lives. The psi-function of the entire system would express this by having in it the living and dead cat (pardon the expression) mixed or smeared out in equal parts.

The key issue here involves the implications of Bohr's point that you have to take into account the entire experimental situation—the interaction between instrument and quantum object. You cannot "drop out" the instrument when measuring a quantum object the way you can a telescope when studying a heavenly body. The quantum object can "be" in many ways, and how it "is" in your laboratory depends on how you set up your equipment. Schrödinger concludes that section of the paper with the observation that there is nothing unclear or contradictory about blurred realities, noting that "There is a difference between a shaky or out-of-focus photograph and a snapshot of clouds and fog banks." Quantum mechanics gives us an analogue to the latter. Blurring in the microscopic world is resolved only when it is observed—that is, made macroscopic. Later in the article, Schrödinger spelled out the implication in the terms that worried Einstein: the nonseparability of previously interacting quantum states even after the interaction, or the violation of what the authors of the EPR paper called the principle of "local reality."

Schrödinger now invented a neologism for the quantum violation of this principle: entanglement.[6] In the classical world, he pointed out, when two bodies interact and separate, our knowledge about the two bodies splits into two pieces, and our knowledge of the two is the sum of what we know about each, thus recovering what Einstein called local reality. In the quantum world, however, when two bodies interact they form a system governed by the ψ-function that does not separate until a measurement occurs. Until then, Schrödinger said, the "entanglement" of our knowledge about them persists.

But it doesn't seem that improbable to me. It seems to me that all the things we keep in sealed boxes *are* both alive and dead until we open the box, that the unobserved is both there and not. Maybe that's why I can't stop thinking about the other Will Grayson's huge eyes in Frenchy's: because he had just rendered the dead-and-alive cat dead. I realize that's why I never put myself in a situation where I really *need* Tiny, and why I followed the rules instead of kissing her when she was available: I chose the closed box. "Okay," I say. I don't look at her. "I think I get it."

—*Will Grayson, Will Grayson*

Schrödinger meant his cat tale as a gag. It carried forward the running joke he and Einstein had been developing: "How silly," he was saying in effect, "to think that the existence of something like a *cat* could possibly be indeterminate, so that its reality depends on whether we observe it!"

Science historian Stephen Brush has remarked that the cat example "captures the spirit of Einstein's own critique better than the published EPR paper."[7] Yet when Schrödinger's paper came out, the cat provoked little discussion. Why should it? It was a silly image for something that everyone thought was a nonproblem. For Bohr, Heisenberg, and company, cats are too complicated to have ψ-functions. They inhabit classical territory. Reality is not blurred. Cats are dead or alive. Schrödinger's image illegitimately postulated an object on one side of the classical-quantum border behaving according to laws that reign on the other side. The example is—to use the slang of the teenagers in *Will Grayson, Will Grayson*—a fail.

For Einstein and Schrödinger, on the other hand, the image

merely showed that, just as we are not content to accept a "blurred model" to represent reality in the macroworld, so we shouldn't in the microworld. Dogmatic q.m. is a fail. Things are different on the other side of the boundary, but not *that* different. For both sides, the cat symbolized nonsense.

For this reason not many references to the cat surface in the next few decades; for a while the cat appears only in letters between Schrödinger and Einstein, as a convenient symbol for what they thought was wrong with the Bohr-Heisenberg interpretation of quantum mechanics.[8] For instance, in 1950 Einstein wrote a letter to Schrödinger repeating that in a complete description of the cat this creature is "real"—meaning dead or alive. Einstein's position, again, is what philosophers call commonsense realism; the cat has the property of life or death independently of whether or not we see it. Bohr and his Copenhagen cohorts argued that such realism is impossible to maintain in the microworld, and the cat cannot cross the boundary into that territory. For both sides of the issue, the cat was a symbol for a problem that you had only if you made the mistake of misconstruing the classical-quantum boundary. "Schrödinger's Cat is a red herring," as the joke goes.

The boundary dispute, however, did not vanish. It grew worse. To the consternation of Einstein and Schrödinger, no way was found to reformulate the rules of the quantum domain so that "determining parts" could all coexist. They had hoped that underneath the wave function additional degrees of freedom could be found, knowledge of which might restore some degree of causality but the extreme determinism of the wave function left it no wiggle room. Entanglement did not go away, and remained an ontological, not just epistemological, disturbance.

But to the consternation of Bohr and Heisenberg, neither was a way found to pin down the classical-quantum boundary. It is easy to say that the problem will be solved once the classical-quantum boundary is located. But no theory has arisen to say that

the boundary is located here rather than there, and no experiment has managed to detect such a boundary. What makes it difficult to observe is that the more complex the phenomenon at which you are looking, the more random effects arise to wipe out the quantum effects for which you are looking. The boundary dispute persisted, making the cat increasingly the symbol of an active problem.[9] Today, the boundary transition is known to be sometimes abrupt and sometimes gradual, involving a proliferation of many different happenings that causes the superimposed quantum system to lose coherence and turn into a classical object with no possibility of returning to its previous state. Such a gradual passage is called decoherence. One of the principal challenges to quantum computers is to preserve the quantum system as long as possible and stall decoherence.

By the time popular-science writing took off in the 1970s, Schrödinger's cat was no longer a gag. It had become a vivid and accessible image to capture the weirdness of entanglement, superposition, the measurement problem, and the ψ-function—indeed, *the* symbol of the challenge posed to conventional realism by quantum mechanics. It had the additional appeal of using something warm, fuzzy, and familiar to make a rather difficult epistemological point. Popular writers pounced triumphantly on the image, proclaiming that it represented the quantum mechanical picture of *all* reality, rather than illustrating merely properties of certain sets of variables. In *The Dancing Wu Li Masters* (1979), an over-the-top book about alleged connections between quantum mechanics and Eastern mysticism, Gary Zukav wrote, "According to classical physics, we get to know something by observing it. According to quantum mechanics, it isn't there until we do observe it! Therefore, the fate of the cat is not determined until we look inside the box."[10] Quantum mechanics, Zukav continued, demonstrates that the world "is not substantive in the usual sense," adding that "Quantum physics is stranger than science fiction." "Schrödinger's Cat," Zukav concluded, "long has illustrated to physics students the psychedelic aspects of quantum mechanics." Zukav's book was wildly popular and won an American Book Award. Its runaway success helped to spread familiarity with the cat image in the explosion of popular physics writing that crested in the 1980s. The cat appeared more often not only in science fiction but also in fringe and mystical fiction, amateur philosophy, and self-help literature.[11]

Why? First, the image can serve to stand in for a certain feature about everyday reality that one might call ontological uncertainty—that while several options may be open to you, once you choose one the others become unavailable. When you choose to marry one spouse you lose the possibility of having a life with others. When you pursue one career others become impossible, at least without considerable retreading. Secondly, the image involves a cat. Schrödinger's cat is not only a convenient metaphor with scientific cachet, but also picturable in an irresistibly cute way. Thirdly, it involves an amusing absurdity, like the Zen koan of the sound of one hand clapping. We love paradoxes. Supposedly they make us think but in reality they tell us we don't have to bother.

Physicists do not seem to care much about Schrödinger's cat anymore except as a label: "cat-states" is sometimes used to refer to large coherent quantum systems, though nothing nearly as complex as a cat. The rest of the world, however, seems to care quite a bit. Some scientists tried crafting careful popular explanations of the cat, including Heinz Pagels in *The Cosmic Code*, Martin Gardner in an essay entitled "Quantum Weirdness," and John Gribbin in *In Search of Schrödinger's Cat*. Truth be told, these sober versions were only marginally less weird than the more breathless ones. Moreover, popular discourse, as we've seen, does not necessarily respect technical explanations, and often adopts and transforms scientific terms for its own purposes. In much the same way that "quantum leap" came to be applied to discontinuous transitions of all sorts, even and especially large ones, so "Schrödinger's cat" and its associated equation came to represent any process involving an indefiniteness that is cured, clarified, or altered by observation or encounters with the world. It's become a metaphor, too, for an ontological distance between one reality and another.

Three-quarters of a century later, Schrödinger's cat is almost as much of a pop icon as Dr. Seuss's Cat in the Hat. Our students are always finding new and inventive references. Some are silly, crude, and on the sophomoric side, involving animal abuse or the cat's imagined revenge fantasies. Others are witty cartoons that twist readers' expectations.

The physics behind the cat paradox has spawned its own genre of jokes. The one that made us laugh the loudest began with two people—let's call them Alice and Bob—flirting in a smoky bar. They get more and more intimate before finally—and as we intend this book for a general audience you'll have to supply the explicit content yourself—seeming to perform two incompatible sexual acts simultaneously. This puzzles the barman, who in the haze cannot make out exactly what they are doing. "What's going on?" he says to the house drunk. "I can't quite see it—it looks wonderful but it doesn't make any sense." "Yeah," the drunk sighs wistfully, "it's a super position."[12]

Getting back to the cat, this one's humor is funny in a hard-to-explain way:

Other mentions of the cat are somber, alluding to darker questions about reality. In the novel *I Married You for Happiness*, by Lily Tuck, Philip, the husband, is a mathematician and Nina, the wife, a painter. Schrödinger's equation, and his cat, keep recurring as metaphors for the gap between the two spouses: abstract and concrete, cerebral and emotional, logical and imaginative, mathematical and artistic, intellectual and practical, eternal and ephemeral. Philip has knowledge of an ideal world that Nina wishes she shared and of which she is vaguely envious. She finds her husband serene and detached, and though she thinks she is more in touch with the real world she finds her life disconnected and discontinuous, her thoughts and memories often hopping around and distracted by trivial things. Schrödinger's wave function and its collapse, the phantom and real cat, are often invoked to mirror the distance between these two people, very close but also irremediably far apart:

> Our brains—how often has Philip tried to explain this?—cannot function in the world of quantum uncertainty. Quantum mechanics is a mathematical construct that embraces two incompatible alternatives, assigning to each its probability.
> Only if we accept an interpretation of quantum mechanics, he goes on to say—but she has stopped listening and is thinking

about something else: How to mix her paints to get the right carmine red? How long to keep the *daube de boeuf* in the oven?[13]

The cat image is found so frequently in popular culture that blowback was inevitable. Here's one example:

Schrödinger's Cat.   icanbarelydraw.com   CC BY-NC-ND 3.0

Another protestor can be found on YouTube in a video called "Schrödinger's Cat—The World's Worst Metaphor," posted by a German teacher going under the name of Sillysparrowness.[14] While Sillysparrowness claims to be a "humanist" and "literary scholar," she professes a strictly literalist interpretation of meaning. Schrödinger's cat image is a *"bad* example, *rubbish* metaphor" because "you just can't explain an event on an atomic scale on a macroscopic scale—it *doesn't work properly."* This is not a person that you want to study French symbolist poetry with.

The cat image continues to fascinate because we find the world flooded with plural, mixed, and incomplete identities—abstract and concrete, East and West, Christian and non-Christian, hetero- and homosexual, male and female, etc.—which persist in a confused way if disengaged from the world, but which can morph into specific and concrete forms in specific situations. We know what it's like to hold ourselves back from the world, and then have irrevocable changes occur when we make contact with it. We know what it's like to juggle multiple identities, and then be forced to choose. These experiences are hard to describe. Most of the conventional forms of expression feel inadequate.

Teenagers know this. This is why the protagonists in *Will Grayson, Will Grayson* (there are two different ones with the same name) do not treat Schrödinger's cat as a gag, and discuss it with some urgency. They experience mixed identities that then abruptly crystallize all the time; for instance, one Will finds his life changed by an encounter in an adult video store called Frenchy's, and the other Will realizes that he holds himself aloof both from his friend Tiny, who is gay, and from the girl Jane, who is attracted to him. The "entanglement" issue surrounding the cat is a great metaphor to express the reality of their mixed feelings, conflicting identities, and unexpressed passions.

The novel shows us more than teenagers making sense of their experience. We see something bigger—*how* they make sense of their experience. Why do these teenagers—not science students—talk about the cat image? Because they cannot understand their lives, they look to ideas for assistance, nothing conventional comes, they feel stuck, and the image seems to help. The cat gives them an example of the kind of thing they feel is happening to them. They use the image to make sense of their relationships, even though they know they're not subatomic particles. Their interest is not quantum physics and the subatomic realm, it's themselves and their own world. They are confused teenagers who live in a world of dualities, of mixed and unexpressed thoughts: smart and not smart, attractive and not attractive, longing and fear. The cat image—with its scientific and cerebral cachet—allows them safely to think that it's cool to think deeply about subjects in a way that also permits them to make sense of what is happening to them, and their own feelings. The protagonists in *Will Grayson, Will Grayson* are using the image in a sophisticated way, to distance themselves from themselves, and view themselves abstractly, but in a way that gives them a better grip on themselves. Talking about the cat image is a comfortable surrogate for talking about themselves, which they can then give up once it's safe.

Once again—as for all of the quantum images we discuss in this book—this is a new image for an old issue. Surrogate speech is not just for teenagers. Another famous example in literature is in Tolstoy's *Anna Karenina*, when Levin, looking into the "kind but frightened eyes" of Kitty, the woman whom he loves but who rejected him in the past, engages her in a word game in which they draw letters with a piece of chalk, which allows both of them to talk safely about the volatile subject of that previous rejection, working through it to reach a happy conclusion.

The cat image in *Will Grayson, Will Grayson* is a tool much like that word game. It's disarming to the participants, but for a different reason in each case: because it's cool and scientific to the teenagers, and playful to the Russian aristocrats. We see that ideas such as Schrödinger's cat are not frozen and rigid, but able to be used as tools in other situations, addressing enigmatic parts of experience.

"Well, that's not all, actually. It turns out to be somewhat more complicated."

"I don't think I'm smart enough to handle more complicated," I say.

"Don't underestimate yourself," she says.

The porch swing creaks as I try to think everything through. I look over at her.

"Eventually, they figured out that keeping the box closed doesn't actually keep the cat alive-and-dead. Even if you don't observe the cat in whatever state it's in, the air in the box does. So keeping the box closed just keeps *you* in the dark, not the universe."

"Got it," I say. "But failing to open the box doesn't *kill* the cat." We aren't talking about physics anymore.

"No," she says. "The cat was already dead—or alive, as the case may be."

"Well, the cat has a boyfriend," I say.

"Maybe the physicist likes that the cat has a boyfriend."

"Possible," I say.
"Friends," she says.
"Friends," I say. We shake on it.

—*Will Grayson, Will Grayson*

In science fiction such as *Blueprints of the Afterlife*, the cat story makes plausible weird and otherwise magical plots—the experience of seeing actual alter egos. For science writers, it expresses succinctly the problems with commonsense realism. For those inquiring into how quantum mechanics is connected with the everyday world, it captures the idea of an "intermediate level of reality" that Heisenberg said was the price we had to pay for quantum mechanics. For geeky teenagers, talking about it is absorbing in and of itself—and just happens to allow them to share feelings that deepen friendships. It only becomes fruitloopery—wacky and pretentious—when its metaphorical origin is forgotten and its meaning is considered to come from its origin rather than how it's used, as if its scientific origin gave it special, magical powers.

The Border War

One of the longest and most intractable border wars ever fought has been over the boundary between the classical and quantum realms.

The first attempt to characterize the boundary was Bohr's "correspondence principle," according to which as quantities with the dimensions of action (the units of Planck's constant h) become large compared to h, a classical description of the motion will become more and more accurate. That's sort of like characterizing the border between the United States and Canada by saying that the farther north you go, the more Canadian things become. It does not pin down an exact boundary—it suggests that you can't do so—but does say that the boundary involves a smooth transition from one place to the next.

Bohr first applied this principle (without the name) in his work on the Bohr atom. For very large orbits, he argued, one should be able to deduce the radiation of light from an atom by using classical physics, and then matching the classical quantities with their quantum counterparts. This enabled him to derive rather than just assume an idea already suggested by others, that angular momentum should be quantized in units of $\hbar = h/2\pi$. As he might have put it, in the limit of large quantum numbers the quantum treatment should go over into a classical one. As you turn the crank to make h smaller and smaller you get more and more Canadian, until at zero you are fully there! Because the energies of atomic levels get closer and closer together, the abrupt jumps or leaps that Bohr invented become smaller and smaller and begin to blur into each other. The point is that the transition between the quantum and classical descriptions is completely smooth.

Though the orbits are large and the associated energy differences between levels are very small, this system of one electron circling around a nucleus still has a small number of components. What happens when the number of pieces in the system becomes very large; what's the biggest object you could get? The subject, of course, invites jokes. "Socks," wrote one *New Scientist* reader authoritatively. "In my experience it can be at least as big as a sock, as they seem to pop out of existence at random (check the number of odd ones in the sock drawer)." The correspondent added, "I have discovered they can be prevented from disappearing by entangling before exposing them to high temperatures in an aquatic environment."[15]

Scientifically, the issue is complicated, for we have to consider not only the number of pieces in the material but also whether these pieces can move independently. For example, in a large perfect crystal at very low temperature there are not many excitations possible, and in principle one could imagine treating the crystal as a single object obeying quantum rules. In practice we do not yet have the technology to do this, but characteristic wave diffraction patterns have been observed by passing buckyballs—molecules shaped like balls with sixty or seventy carbon atoms in a cage-like structure—through an appropriately designed grating. Classical particle physics also would give oscillating intensity patterns, though very different ones, simply because of the periodic nature of the grating, but the results turn out to look like those expected from the quantum waves.

What about the diffractions of much larger objects, in which case the spacing of a diffraction grating would have to become correspondingly large? Here the first problem would be the thermal excitations that would make the beam too incoherent to interfere properly. To take an absurd example, if you wanted to diffract people you would first have to cool them to such a low temperature that they could not be revived!

The example of a large crystal suggests that one might be able to detect quantum interference in objects that are big enough to be seen in a microscope, and a couple of experiments have achieved this, one by our colleagues at Stony Brook.[16] In that experiment, a superconducting loop has external magnetic flux through it which is halfway between two superconductor quanta of flux. Thus the superconductor has to supply a current that brings the flux down half a quantum or up half a quantum, and these two configurations have very close to the same energy, so that one can look for quantum oscillations between them. These quantum oscillations were actually detected, meaning quantum interference between two states differing in magnetic moment by about 10 billion times the magnetic moment of a hydrogen atom! That is getting pretty macroscopic.

One of the early proponents of doing such experiments was the Nobel laureate (for theoretical work on helium-three superfluids) Tony Leggett, who was skeptical that they actually could be done. His skepticism about further extensions of quantum interference into the macroscopic world continues, as revealed dramatically in a debate about whether quantum mechanics in its current form can account for our perception of definite outcomes (e.g., something as complex as a cat in a box always seen as alive or dead, not in a superposition of the two states) held at the 2005 symposium "Amazing Light" in honor of Charles Townes's ninetieth birthday. Leggett's opponent in the debate was Norman Ramsey, himself a Nobel laureate for work on interference systems used, among other things, for making atomic clocks. Ramsey was older, and one might have thought therefore more inclined to favor classical over quantum, but he took the position that there is no intrinsic limit to the size or complexity of objects which can exhibit quantum interference.

In practice, what makes interference disappear as systems get larger? The majority view is that as size increases there can be more and more "loose" degrees of freedom coupled to whatever is the main system. If these degrees of freedom are in different states corresponding to certain states of the main system, then it becomes extremely hard for those states to interfere, because not only do they have to feed into a single channel, but so do the extra states. Thus the presence of such "quantum fuzz" can cause decoherence, meaning the inability of different macroscopic states to interfere with each other.

Schrödinger's cat is a perfect example. The two states "alive" and "dead" are extremely messy combinations of many different variables that automatically have quite different values for the two states. To mention just one, there is the number of hemoglobin cells carrying carbon dioxide versus the number carrying oxygen. For full interference those numbers should be identical for both states, but that is highly unlikely.

Leggett was exploring, among other possibilities, an idea originally proposed by others that there may be new—so far undetected—types of, fluctuations that couple more and more easily to large systems, and thus make observable interference increasingly difficult, no matter how good the experimenter's control of the known degrees of freedom. It's certainly worth looking for such a phenomenon, as it would add a significant new feature to our knowledge of the physical world. Fortunately this searching can be done simply by trying to make larger and larger systems that do exhibit interference. If Leggett is right, this would be like discovering a "natural barrier" to increased size and complexity of systems exhibiting coherence, the equivalent to a natural boundary between the United States and Canada.

Meanwhile, experimenters and theorists continue to put objects into ever bigger quantum superpositions. Recently, two physicists at the University of Duisburg–Essen attempted to quantify the size or "macroscopicity" of experiments with such objects. They called their newly defined paramater μ, of course—the Greek letter pronounced "meew." The larger the μ-value, the more macroscopic the object. The researchers calculated that the most macroscopic superposition so far obtained, in an experiment they helped conduct in 2010 with 356 atoms, had a μ of 12. They thought it might be possible eventually to conduct experiments that would raise μ to 23. But a cat—which they modeled as a 4 kg sphere of water, among other of a set of vast simplifications—seems hopelessly out of reach at μ = 57, or equivalent to what it would be for an electron to exist in a superposed state for 10^{39} times the age of the universe. "One should never say never," one of the two researchers noted, "but we will probably never be able to put a cat in a quantum superposition." Theorists pointed to other difficulties, serious and frivolous, including the problem of herding thought cats. "[I]t all got a bit silly," the editors of *Physics World*, who reported the story, remarked.[17]

The distinctive feature of the quantum realm, the feature that is ultimately responsible for every bit of its weirdness, involves wave properties giving predictions for particles. Entanglement, superposition, polarization, interference are familiar for classical waves, but absurd for classical particles. When the border between the quantum and classical realms is finally fixed, we can be sure of one thing: waves and particles go their own ways at the checkpoint.

Chapter Eleven

Rabbit Hole:
The Thirst for Parallel Worlds

In David Lindsay-Abaire's award-winning play *Rabbit Hole* (2005), Jason, a high-school student, was driving down a quiet, suburban block when a dog suddenly dashed out in front of the car. Jason swerved, but struck and killed a boy chasing the dog. Grieving and remorseful about this tragic and unnecessary accident, he writes a story for his school literary magazine. It is about a scientist who finds a network of passageways to parallel universes with different versions of people in this one. After the scientist dies, his son explores those different worlds, hunting for a universe in which his father is alive. Jason gives a copy of the story to Becca, the mother of the boy he killed.

Jason's vision of traveling to parallel worlds in *Rabbit Hole*, which was made into a film starring Nicole Kidman (2010), becomes tangible in *Another Earth*, a film with a similar theme. The protagonist this time is Rhoda (Brit Marling), a high-school senior whose life is rich in possibilities after she gets accepted to the Massachusetts Institute of Technology, where she can pursue her passion for astrophysics. After a party celebrating with her friends, she smashes her car into another, sending its driver, John, into a coma and killing his pregnant wife and child.

Rhoda is next seen moving back in with her parents after being released from a four-year prison sentence. Meanwhile, an earth has appeared in the skies, identical to ours except that events in the lives of its inhabitants branched off at the time of its discovery—which happens to be just before Rhoda's accident. A company, United Space Ventures, begins booking trips to what's being called Earth II. Rhoda, still traumatized after her jail term, wins a ticket, hoping to visit her alter self with an unruined, no doubt fulfilled and productive, life. At the last moment, she changes her mind and gives the ticket to John, the man whose family died in the crash, allowing him to see them again, as an intact family.

The idea promulgated in both *Rabbit Hole* and *Another Earth* of alternate worlds is not new. It has been around since ancient times—although not necessarily as a purely scientific idea—and has served many imaginative purposes. In the fifth century B.C., the Greek philosopher Philolaus proposed the existence of a "counter-earth" or *autochthon* that remains invisible to people living in our hemisphere. The purpose here is cosmological; without the counterearth Philolaus's universe would have been imbalanced. Another difference is that Philolaus did not envision the counterearth to have counterparts to us. In classic Greek mythology, meanwhile, the myth of Orpheus envisions the possibility of bringing someone back to life—thus, reversing an unfortunate death—through love and dedication.

Jumping to the nineteenth century, Lewis Carroll's beloved story *Alice's Adventures in Wonderland* uses an uncanny, fantasy world for the purpose of entertainment and satire. *Men Like Gods*, a 1922 story by H. G. Wells, uses an alternate world for social criticism. In it, a journalist is transported to Utopia, a parallel world that was once much like earth but is now a much advanced version of 1920s England, lacking many of its maladies, including government. (As the novel's hero Mr. Barnstaple is told, in Utopia "Our education is our government.")

In his 1941 short story "The Garden of Forking Paths," published in a collection of the same name, the Argentine writer J. L. Borges—a relentless explorer of philosophical paradoxes involving time and identity—envisioned a "multiverse" in which all possibilities, emerging from all possible decisions, coexist. "The Library of Babel," another story in the book, is a literary counterpart, in which Borges depicts a gigantic library containing every possible permutation of letters, spaces, and punctuation of 410-paged books of a certain format, in order to explore puzzles involving meaning and information.

The idea of alternate realities, or the fact that our lives or our world would have been different had we made different choices, is also not new. Robert Frost's poem "The Road Not Taken" (1920) ponders the choice of a horseback rider between two paths home. "Two roads diverged in a yellow wood," Frost writes, leaving him sorry he could not travel both. In *Intimate Exchanges*, a series of plays by the English playwright Alan Ayckbourn, the opening scene branches out into sixteen different endings, depending on insignificant choices that the protagonists make, with two versions presented per evening over eight performances. It is a theatrical tour de force illustrating, as Ayckbourn once wrote, "those tiny decisions we all make in our lives that lead to bigger consequences."[1]

What's different about *Rabbit Hole* and *Another Earth* is that they exploit a very specific form of alternate reality, consisting of multiple parallel worlds almost identical to and simultaneous with ours, which have branched off from each other due to split-second decisions that were not consciously taken. This idea was born in the unlikeliest of places: a highly controversial interpretation of quantum mechanics. To keep things straight, we distinguish between *parallel* worlds, or the idea that many equally real worlds coexist, and *alternate* worlds, or the idea that the existing world is the only real one but might have been otherwise had different paths been followed. The idea of parallel worlds is yet another way in which quantum mechanics has contributed a new focus to ancient dreams and desires.

Everett's Folly

Quantum mechanics, we said at the beginning of the last chapter, describes the subatomic world as unfolding due to the intersection of two ingredients: Schrödinger's equation, which contains many superposed possibilities, and something that makes one of those possibilities appear. To connect this strange picture with our familiar, everyday world, physicists adopt one of three general strategies.

The first and most common is called the Copenhagen Interpretation, named after the city in which Niels Bohr was based. It divides the world into two very different domains: the quantum and the classical. The quantum domain is governed by a field that is not itself tangible or observable, and is described by Schrödinger's equation—a recipe giving the probabilities of certain real states' materializing. When this quantum field encounters something in the classical domain, through a measurement or other interaction, the function evaporates or "collapses" in the encounter, eliminating all but one real state. All we can ultimately know of the world to come is probabilities.

The Copenhagen Interpretation is itself sufficiently strange that it has inspired two sets of deniers. One set, discussed in Chapter 9, denies the first ingredient. There is, these deniers say, no true superposition or information wave function; our knowledge of the quantum field is incomplete. More factors exist out there than we have discovered, and once we discover these "hidden variables," causality and predictability will be restored. The other set of deniers rejects the second ingredient. They argue that we do not—and cannot—eliminate the other possibilities: they *all* exist.

This is the parallel or many-worlds interpretation of quantum mechanics, one of the most logical, bizarre, and ridiculed ideas in the history of human thought, and the inspiration for many science-fiction stories. The many-worlds idea was introduced by Hugh Everett III. Everett (1930–1982) led a short and fragmented life.[2] Born in Washington, DC, he was a compulsive model-builder. He "burned to reduce the complexity of the universe to rational formulae."[3] Yet as he tried to grasp life through models he kept losing track of his own.

Everett's personal life was as erratic as his career. He was a stubborn, overweight, chain-smoking alcoholic who ignored his children and mistreated his wife, whom he married in 1956, and was a persistent womanizer. In his affairs with women, says a friend, "his objective function didn't include emotional values." Says another friend, "He looked at life as a game, and his object was to maximize *fun*. He thought physics was fun. He thought nuclear war was fun." Modeling it, anyway.

Everett was too obsessed with models to intersect effectively with the real world. At the end of his life, near bankruptcy, he tried to write code for a software program called "winning mortgage" to calculate mortgage payments in various scenarios. His daughter, a manic depressive who married a drug addict and became addicted herself, committed suicide. Everett died of a heart attack while drunk. As paramedics carried the corpse away, his son realized that he did not ever remember having touched his father in life. Following his wishes, Everett's widow tossed out his cremated remains in the trash.

Everett entered Princeton in 1953 to study physics, and became intrigued by quantum mechanics and its "measurement problem," a common name for the way that the superposition described by the wave equation "collapses" or evaporates when a measurement event takes place. Everett was dissatisfied by the way that the Copenhagen Interpretation casts the universe as a cosmic apartheid, divided into a determinate, real domain where measurers live and where measuring takes place, and an indeterminate, unreal quantum domain. He found it awkward that the wave function that rules the quantum domain evolves continuously, while the measurement act that intrudes from the classical side is abrupt and discontinuous, magically eliminating all possibilities except one.

In Everett's doctoral dissertation (1957), written under the supervision of the physicist John Wheeler, he found a way to eliminate this puzzling situation that is as simple as it is outlandish. When a quantum system is measured, or otherwise interacts with the classical world, the superimposed possibilities don't vanish—the system splits into parallel worlds, inhabited by almost-identical twins of our own. Each of these worlds itself keeps branching, bushlike, the junctures being every place where the quantum domain contacts the classical world. In Everett's words:

> with each succeeding observation (or interaction), the observer state "branches" into a number of different states. Each branch represents a different outcome of the measurement and the corresponding eigenstate for the object-system state. All branches exist simultaneously in the superposition after any given sequence of observations.

The observer state, it seems, not only travels both of Frost's roads but all other possible ones, simultaneously.

Everett submitted a short form of his dissertation to the *Reviews of Modern Physics* for a special issue. The editor of that issue, Bryce DeWitt, found Everett's work "beautifully consistent" but implausible, saying, "I simply do not branch." DeWitt wrote to Everett that he had not addressed the key perplexing issue of quantum mechanics, namely, the transition between the possibilities described by the wave function and the actual situation that comes to be in and through the measurement. In response, Everett added a footnote to his paper, saying that his theory takes care of this issue in "a very simple way," namely, that in it "there is no such transition." The footnote continued:

> From the viewpoint of the theory all elements of a superposition (all "branches") are "actual," none any more "real" than the rest. It is unnecessary to suppose that all but one are somehow destroyed, since all the separate elements of a superposition individually obey the wave equation with complete indifference to the presence or absence ("actuality" or not) of any other elements. This total lack of effect of one branch on another also implies that no observer will ever be aware of any "splitting" process.

Really, skeptics asked, we won't ever be aware of the splitting? Right! Everett emphasized in the footnote:

> Arguments that the world picture presented by this theory is contradicted by experience, because we are unaware of any branching process, are like the criticism of the Copernican theory that the mobility of the earth as a real physical fact is incompatible with the common sense interpretation of nature because we feel no such motion. In both cases the argument fails when it is shown that the theory itself predicts that our experience will be what it in fact is.[4]

DeWitt later came to be a champion of Everett's theory, and with his student Neill Graham coedited a collection of papers on it. They point out that "Everett's interpretation calls for a return to naïve realism and the old fashioned idea that there can be a direct correspondence between formalism and reality."[5] This is not entirely correct. The only conceivable way to "see" the alternate worlds so that the wave function formalism corresponds to them would be to stand outside them, having a God's-eye view with an intuition of these worlds without "observing" any. If a human being observes any of them, that human being is a part of that specific world and is unable to observe the others. In Schrödinger's cat image, the only way for there to be two coexisting possibilities is for someone outside the box. In the universe, however, we are all in the cat's position. From the cat's view, there is only one reality.

Furthermore, the wave function is an instrument of prediction. But the function only states the best information I have from *this point* about what will happen next to the system. Schrödinger's equation is a projection toward the future; for there to be super-position, you have to be on the outside or at the beginning. As a calculational tool, it does not mean that the system really is in all those states, only the probability that each one will become real. It doesn't tell you if the cat is dead or alive; it's not made to. In the double-slit experiment mentioned in the Interlude to Chapter 8, for instance, the wave equation tells you not whether the electron will be here or there, but the probabilities that the electron will be here or there. The vanishing of the probabilities is no more strange than the vanishing of the probabilities that I, or anyone else, will win the lottery once a winner is chosen. Everett's idea, on the other hand, gives reality to all the superposed possibilities. The idea's a cheat, but survives because it also assumes a cloak of invisibility—that no state can see another.

Another odd feature about Everett's idea is that because the alternatives are amplitudes, they are not occurring in a different geometrical location but all in the same space-time. The parallel worlds are not spatially separated; in a measurement, all the pos-sibilities are happening at once in the same place. The cat-box is crowded; they are dead and alive at once. But how can that be, for the dead cat is presumably lying down and the live one sitting up?

Einstein's approach was to say that quantum mechanics does not fit our most naïve view of reality. Everett's approach was to rede-fine reality in a garish way—as everything that might be happen-ing. Everett's interpretation restores causality and eliminates the collapse of the wave function by making hash of the idea of reality itself; it postulates an infinite number of infinitely branching "real" worlds that are "real" only to a divine intuition.

"[N]o one could fault his logic, even if they couldn't stomach his conclusions," wrote Charles W. Misner, a fellow Princeton graduate student. "The most common reaction to this dilemma was just to ignore Hugh's work."[6] Everett left the field and never published again about quantum mechanics.

Fortunately, the Cold War created a huge market for game players and modelers, used in military research to chart outcomes of strategies for conducting nuclear war. Here Everett found more respect, having invented an "Everett Algorithm" to improve on the traditional Lagrange multiplier method for calculating consequences in logistics problems. In the 1950s and 1960s, he worked for the Pentagon's top-secret Weapons Systems Evaluation Group, devising strategies for waging nuclear war and estimating the lethal effects of fallout, and for the Lambda Corporation, another military think tank.

"Schizophrenia with a Vengeance"

DeWitt, Everett's first editor and an initial skeptic, soon changed his mind. In 1970, wanting to attract attention to Everett's take on quantum mechanics, DeWitt wrote what he later called a "deliberately sensational" account of it in *Physics Today*, renaming it the "many worlds" interpretation. DeWitt's vivid prose certainly woke people up: "[E]very quantum transition taking place on every star, in every galaxy, in every remote corner of the universe is splitting our local world on earth into myriads of copies of itself. . . . Here is schizophrenia with a vengeance."[7] Still, most physicists found Everett's idea—that one macroscopic wave function exists for the entire universe, with a different coordinate variable for

Nature, July 5, 2007.

every particle, continually bifurcating into versions with different probabilities—logical but unhelpful.

Nature magazine captured the attitude of most scientists well on its cover of July 5, 2007, which was done up garishly in imitation of a trashy, 1950s science-fiction tabloid. *Nature: Astounding Tales of Superscience* depicted a woman about to scream, looking over her shoulder at an endless line of near-identical twins—though it was an indispensable part of Everett's notion that the woman could have no communication or knowledge of her twins. The feature story was entitled "Many Worlds: Fifty Years of the Ultimate Quantum Strangeness."

For most scientists, the problems with Everett's theory are that it creates no new predictions, assumes that the bifurcations preserve entire worlds, and eliminates one problem at the cost of postulating the existence of myriad worlds that we cannot detect. In a garish violation of Occam's razor—the principle that the most economical explanation is the best—it makes the idea of "reality" all but meaningless. Why postulate uncountable infinities of unknowable, branching universes to address a problem for which there are solutions that prune the branches? It's the viewpoint, that is, of someone who stands outside the world, and the world starts moving on without that person's being able to see what's happening; after a certain period of time that person asks, "What might I see?" and the wave equation says, "This is what you might see!" But it's only a set of possibilities, not reality—and there is no such person.

Everett's idea is *merely* an interpretation; it fails to make predictions and is incapable of falsification. Like Schrödinger's cat, it is all but ignored by many physicists (though scientists prone to viewing the universe as a whole from a distance, such as cosmologists, tend to like it), but it is a familiar feature in popular culture.

Fictionally Enticing

In *Slaughterhouse-Five*, Kurt Vonnegut attributed a similar outside-the-box intuition to the creatures who live on the planet Tralfamadore:

> The most important thing I learned on Tralfamadore was that when a person dies he only appears to die. He is still very much alive in the past, so it is very silly for people to cry at his funeral. All moments, past, present and future, always have existed, always will exist. The Tralfamadorians can look at all the different moments just that way we can look at a stretch of the Rocky Mountains, for instance. They can see how permanent all the moments are, and they can look at any moment that interests them. It is just an illusion we have here on Earth that one moment follows another one, like beads on a string, and that once a moment is gone it is gone forever.
>
> When a Tralfamadorian sees a corpse, all he thinks is that the dead person is in bad condition in the particular moment, but that the same person is just fine in plenty of other moments. Now, when I myself hear that somebody is dead, I simply shrug

and say what the Tralfamadorians say about dead people, which is "So it goes."[8]

The British television show *Doctor Who*, the longest-running science-fiction show ever made, included characters called Time Lords who not only stood outside the box but could move anywhere inside it:

> MICKEY: I've seen it in comics. People are popping from one alternative world to another. It's easy.
> THE DOCTOR: Not in the real world. Used to be easy. When the Time Lords kept their eye on things. You could pop between realities, home in time for tea. Then they died. And took it all with them. The walls of reality closed. The world was sealed. And everything became a bit less kind.[9]

The scientific attention provided by Everett and then DeWitt boosted the popularity of alternate worlds, and gave it new possibilities. The idea was a godsend to novelists, moviemakers, and storytellers of all types. If you cheat on Everett's idea by imagining information exchange or travel between alternate or parallel worlds, it opens up a tremendous "plot space." The traditional idea of alternate worlds usually involves two worlds, with things different due to a single decision or changed turning point. It's like time travel in traditional science fiction. The idea of parallel worlds, plus the (scientifically disallowed, in Everett's theory) idea of being able to cross laterally or horizontally (so to speak) between them, instead of back and forth in linear time, provides a new narrative degree of freedom allowing writers, filmmakers, and other spinners of narratives to develop new explorations of puzzles of time, identity, and historicity.

The first story based on the idea, "Store of the Worlds," appeared in *Playboy* in 1959. Penned by science-fiction author Robert Sheckley, it was about an illicit store whose proprietor uses a combination of injections and gadgets to send customers to alternate worlds, cast off by our own, in which they could fulfil secret desires. Disappointingly enough, for the typical reader of that magazine, there is no sex in the story, only a search for an ordinary, normal world by someone from a destroyed one.

In comparison with later treatments of parallel worlds, Sheckley's plot is primitive. More elaborate plots exploiting more complex possibilities soon appeared. Larry Niven's short story "All the Myriad Ways" (1971) conveys the message that travel to alternate universes leads to moral nihilism. A police detective, puzzled by a crime wave following the discovery of travel to parallel worlds, finds that discovery of the possibility has taken all meaning out of moral choices, promoting the idea that any act, even a criminal one, is as good as any other. It's a nihilistic message also embodied in the following comic strip:

And the following cartoon:

"David believes in multiple universes—all of them lousy."

A more action-oriented approach is taken in the movie *The One*, starring Jet Li (2001), whose plot involves the ability to become all-powerful. There are a finite number of universes and therefore versions of each person. When one version dies, the versions in the remaining universes get more power. Realizing this, a multiverse police force has been established to try to keep people from killing their alternates, and to track people through the worlds. A renegade from the police, however, comes close to omnipotence by killing (almost) all his versions.

Neal Stephenson's novel *Anathem* (2008) is a sophisticated example of science fiction depicting parallel worlds. The plot involves an encounter between inhabitants of parallel worlds that respects that their inhabitants may be slightly different even in biological ways, for example, the food that they can digest.

Everett's idea has also been exploited in several TV shows, including the animated comedy series *Family Guy*, about a dysfunctional family called the Griffins. In an episode several years ago entitled "Road to the Multiverse," the Griffins's baby Stewie and dog Brian use a remote control to tour parallel universes, ending up in one in which dogs rule and humans obey. Brian is reluctant to leave.

The idea of parallel worlds has even inspired some sculptors, such as Jon Ardern and Anab Jain of the Superflux studio in London. They created a work called *The 5th Dimensional Camera* (2010), which was part of a 2011 exhibition called "Talk to Me" at the Museum of Modern Art in New York, and featured objects that interwove technology and communication. Their exhibit resembles a camera, mounted on a tripod, attached to a red bullhorn-like cone. The cone is pointed at an array of snapshots arranged in rows and columns on the wall. The snapshots were taken at the same location, and each bears the same date and time—25-11-2012, 15:09:38—but something slightly different appears in each: bike riders, smokers, arguing couples, homeless people, no people at all.

The label reads:

> According to the many-worlds theory, posited by physicist Hugh Everett in 1957, despite the fact that individuals observe time as linear, there is an unfathomable number of universes, each supporting parallel timelines that each have the potential to converse and influence outcomes in the others.

Artistic license: the ground rule of Everett's idea is that each world remains unobserved to, and cannot influence, the others. Never mind. The label continues:

> The 5th dimensional camera, a metaphorical many-lensed object, explores how we might see all those different worlds at the same time. All the possible ramifications of any decision or action or day would theoretically be visible, in all the worlds that branch out from our linear timeline.

The 5th Dimensional Camera is more cerebral than the vicarious pleasures of "Store of the Worlds," the dazzling complexities of *Anathem*, and the slapstick comedy of *Family Guy*'s multiverse episode. It aims more at the intellectual pleasure of puzzling out what it would be like to have technology let us see evolving worlds not our own.

The Attraction

Parallel worlds tap human anxiety about taking the right or wrong path. The idea allows us to entertain, in imagination at least, having it both ways, or the ability to rewrite our pasts. A psychiatrist colleague likes to speak about the powerful impulse he has—and his patients have—to say, "I should have done that other thing!" He refers to the urge to say this and contemplate what would have happened as I-should-have-done-that-other-thing-ing. We find it reassuring even to think that we *could* have made a decision the other way. This gives our identities a twist in a way that (unlike fermions and bosons) *is* conceivable. It hints that there is more to reality than what simply is; it touches and exploits our fear that *this* is the only reality possible. It allows us to feel that we could have done better—that we could have been contenders—and that, in some world, we still are.

Parallel world fruitloopery promises the *actual* ability to contact alternate worlds; you can easily find such promises on the Internet. There do in fact exist parallel worlds, the promise runs, and in one of them you are wildly successful and can learn from that person. You still can be a contender in *this* world, provided that you simply fork over the money for an instructional video.

By contrast, fictional use of the parallel worlds idea helps by allowing us to contemplate the other possibility without fraudulently insisting that it really exists. That this is the only reality— that we had to have made the bad decision—is truly a source of dread. The wistful notion that we could have done otherwise— that we are good but fallen—is more hopeful. Negatively, this is an escapist fantasy—the idea of meeting another you. It's a version of yearning for the Garden of Eden, for a complete, fulfilled reality that we have fallen away from—a suppression of our finitude. Positively, the idea can help us accept our finitude. In *Rabbit Hole*, the idea allows Jason to articulate for himself—and therefore for us spectators—a guilt and remorse that he cannot otherwise express. He daydreams to cope not with the subatomic domain but with his own life. It helps Becca, the mother of the dead child, as well. "Somewhere out there I'm having a good time," she muses.

In *Another Earth*, the parallel world is scientifically implausible: that an identical earth could have formed at all, that it could have remained undetected until now, that people could ever travel between alternate worlds, and so forth. Humanly, however, the use of the cosmic *deus ex machina* is fully convincing, for it helps exhibit and focus Rhoda's remorse and repentance. Rhoda ends up winning a trip to the Other Earth. But—heeding the advice of an old Indian janitor who has blinded himself to "look inside"—Rhoda decides that she would learn nothing by encountering herself. Self-knowledge is exceedingly difficult to obtain, and the temptation is overriding to be distracted and seek wish fulfilment instead. "The most unusual encounter you can have is with yourself," as one of our students put it after watching this movie. In the movie, we see Rhoda tempted by the desire to meet a version of herself that had not made her mistakes—but we also see her realize that this would not help. We see that the prospect of encountering another version of herself heightens Rhoda's sensitivity to her own situation, to her own historicity. It focuses her more on this world than the other. She loses her interest in the wish fulfilment of watching her other, presumably successful self and gives her ticket to John, the man whose life she destroyed, because it has *more* value for him and his family than it has for her. The ticket forces her to refocus what she regrets; the focus moves from her to the situation. What now matters to her is not moving herself forward but moving the entire situation forward—including what has happened to John and his family. This movie with an impossible plot device has shown us a truth—a shift in Rhoda's thinking from "I want to see what I was like" to "I'll allow *him* to see what *his family* was like."

The impossible plot device allows what healing there is, and therefore shows us in the audience something about healing. As Brit Marling, who plays Rhoda and cowrote the script, was quoted in the *New York Times* as saying, "Sometimes in science fiction you can get closer to the truth than if you had followed all the rules."[10]

We live in fragile bubbles—bubbles that keep opening up and closing off possibilities. This is the human truth to which the impossible plot device can sensitize us, one related to our own finitude. Though one of the most implausible and unrealistic ideas in the history of science, the idea of parallel worlds can nevertheless help to heighten our understanding of humanity.

Multiverses

When we discuss parallel worlds in class, some students are puzzled because they confuse that idea with another idea involving other possible universes that has recently made the rounds of science periodicals: the idea of the multiverse. That idea arose as follows: In our current picture of physics there are many parameters for whose values we have no explanation. The idea is that these parameters each could have had a wide range of values, for most of which human life would not be possible. Therefore the fact that we do exist insures that the parameters have suitable values. An example of such a parameter is the ratio of the mass of the nucleus of a hydrogen atom, the proton, to the mass of an electron that is bound to that nucleus to make the atom. Obviously there are many more such parameters. The "multi-universe" claim is that outside our range of observation must be many other universes with very different, randomly determined properties.

It would work this way. Quantum fluctuations in the behavior of what are called strings might lead to forming bubble universes, which start out small and expand. Depending on the way strings happen to lie, the resulting physics parameters become a matter of chance. This produces a so-called "landscape" picture: The quantities that physicists have identified as fundamental constants, such as the charge and mass of an electron, perhaps might be very different in other universes, inaccessibly far away from our visible universe. Indeed, another universe might not even have a particle analogous to the electron at all. Consequently, life as we know it might not be possible in most of these hypothesized universes. This has led to the idea that the reason we find the patterns we do in our universe is that anything very different would have made our life impossible, and so we wouldn't be here, able to find anything! The name given to that idea is the "anthropic principle"— that is, the only universes that can be perceived by humans are the ones in which humans exist.

This idea has not been ridiculed as Everett's idea has, but there are those who find objections to it. The principal objection is that it implies we are never going to be able to explain these properties. That means giving up on the quest of science to go deeper and deeper into understanding the world. There will at some point be a dead end.

This is not the first time there has been such a suggestion. At the end of the nineteenth century there were some who felt that everything significant that could be discovered had been. That turned out to be way off the mark! In view of our subject in this book, we should mention quantum mechanics as one discovery that wasn't on anyone's radar screen at the time. More recently, the science writer John Horgan published *The End of Science* (1996); its title is self-explanatory. This was almost at the same time as the publication by John Barrow and Frank Tipler of a book about the anthropic principle, but before such ideas emerged in string theory. Many scientists have used the principle in a different way, to predict the results of certain experiments based on the notion that these results were necessary to explain how our current universe (and the humans in it) came to be. In this approach, there is not a necessary implication that the results never will be explicable in terms of more fundamental concepts. The string theory landscape, with its enormous number of possible parameter values (presumably to be determined randomly in the evolution of a universe) emerged in 2003 in a paper by theoretical physicist—then at Rutgers and more recently our Stony Brook colleague—Michael

Douglas.[11] This multiplicity of universes in different places is very different from Everett's wave function for the (single) universe, with different components of the wave function corresponding to different possibilities for the universe coexisting at the same time and in the same place.

Coming back to the multiverse idea associated with string theory, one might ask, How could there be quantum oscillations determining whole universes? The answer is given by the string theory picture, in which initially very localized quantum fluctuations expand rapidly to form whole (and still expanding!) universes. Unlike Everett's proposal, which requires coherent fluctuations over the scale of a very large universe, in this case one only finds fluctuations in very small regions, which set the initial conditions that continue to hold for parameters describing the physical laws of each ever-expanding universe.

Chapter Twelve

Saving Physics

O ur visiting speaker to the class one day was David Kaiser,
a science historian at the Massachusetts Institute of Technology. Kaiser's book, *How the Hippies Saved Physics: Science, Counterculture, and the Quantum Revival*, had just been published, and
we had assigned chapters. Kaiser was unable to travel from Cambridge to speak to our Stony Brook class, but generously agreed to
participate in a Q & A session via Skype from his home, after he'd
fed his two young children and put them to bed. After discussing
his book for an hour and preparing questions, our class dialed him
up from the classroom computer.

Kaiser's head suddenly popped up on the huge projection screen
at the front of the room. He was sitting at the desk in his study.
Next to him was a stack of books and papers, in the background
a yellow daybed with a few white pillows. Kaiser is an attractive
young professor and always a hit as a speaker; he has an engaging
voice and animated speaking demeanor; he's always moving, nodding his head, adjusting his glasses, and gesturing with his hands
for emphasis. He is a captivating storyteller who makes you think
that science history is the coolest field of study ever. He briefly
described his book, about the origin of the attempt in the 1970s to
relate the implications of quantum mechanics with ancient Eastern
mysticism and New Age concerns, and offered to answer questions.

To ask their questions, students had to leave their seats and to
sit in front of the class computer off to the side, so Kaiser could
not see or hear the rest of the students or the classroom itself, but
only the student directly in front of the computer. The first stu-

David Kaiser.

dent to do so asked Kaiser about the relation between his approach and that of Paul Forman, the historian who had argued (as we saw in Chapter 4) that cultural anxieties during 1920s Germany were ultimately the reason why the founders of quantum mechanics placed randomness at the core of their theory. Was Kaiser doing the same thing by seeking a connection between the free speech movement, the anti–Vietnam War protests, and student rebellions of the 1960s and 1970s, and the link of quantum mechanics with Eastern mysticism and New Age philosophies?

Kaiser smiled broadly. "Thank you for that question!" he said. "My research is concerned with how and why certain themes and research topics move in and out of the physics mainstream. And I find the 1970s especially interesting because it is a period of rapid change of fortunes." He nodded his head vigorously. "After the Second World War enrollments in physics classes shot up, money and resources for physics were plentiful and kept flowing at a higher and higher rate—that is, until the end of the 1960s. That was the time of protests against the Vietnam War, suspicion of involvement of physicists in defense research, a shift towards applied science— and enrollments declined and resources directed to physics plummeted. In such a period of rapid change, what happens? Now, I had read Forman's work while I was a graduate student . . ."

Suddenly, on the screen above Kaiser's shoulder in the background, a tiny brown paw appeared from beneath a pillow on the daybed. This was visible to everybody in the classroom, but not to Kaiser himself. Nobody said a word. But it was impossible for those of us watching the screen not to switch our attentions back and forth between Kaiser and the paw, in an alternation of foreground and background.

". . . and I found Forman's work both inspiring and puzzling. Forman's work is an ambitious attempt to link ideas and institutions . . ."

A second brown paw now appeared, next and parallel to the first. Gentle titters swept the class, though Kaiser could not hear them. Sixty seconds passed, with Kaiser continuing to speak. A cat suddenly jumped out from behind the pillow on the daybed—evidently the cat had been sleeping—shook itself off, looked for a moment at Kaiser, and then surveyed the room. Most of the class began to giggle.

". . . but Forman has no real mechanism for connecting the two. He has no real 'gears' in his model, where ideas and institutions meet, where there's traction—except for those talks that the scientists give at mostly ceremonial occasions. In Forman's approach the agent is just 'something in the air,' an atmosphere . . ."

The cat now discovered its tail and began to chase it furiously, racing round and round, a miniature whirlwind on the bedspread. The class was now roaring with laughter, the students looking hysterically at each other, some pounding on their desks with delight. Still, Kaiser could neither see nor hear either the cat behind him or the pandemonium in the classroom.

". . . I on the other hand want to look at the specific institutional structure of the time and find those gears. I wanted to look at how people were trained, what they read, where they sat, whom they spoke with—I wanted to find their salons, so to speak, and check out who was hanging out with whom and who paid for what. In that way I thought I could link specific ideas with specific places and institutions. Does that answer your question?"

It did! The next questioner asked Kaiser how he treated his cat, and clued him in to what was happening. Kaiser turned around; the cat, observed, leapt off the bed and vanished from the screen. "Its name is Quizzie!" Kaiser said, laughing good-naturedly. "Short for 'Inquisitive.'"

Quantum Mysticism

How the Hippies Saved Physics is about a collection of California eccentrics called the Fundamental Fysiks Group who, in the 1970s, explored the possibility that quantum mechanics pointed to deep, mystical truths whose only parallel is to be found in Eastern religions, giving birth to a New Age–affiliated movement that can be called quantum mysticism. If Forman's work raises the possibility that the Quantum Moment is not a proper moment at all but a cultural by-product of a particular moment in Western history, quantum mysticism suggests that what we are really talking about should be called the Eastern Moment.

The book's title is cheeky. But it's a cheeky tale, with a moral for scientists and philosophers. Let's review its backstory. The creation of quantum mechanics in 1925–27, as we saw in Chapters 7 and 8, was a stunning scientific achievement that also sparked a genuine philosophical crisis. The theory overturned basic beliefs about space and time, causality and reality, and exposed as mistaken many fundamental cultural and philosophical assumptions about science. The "bottom has dropped clean out" of the world, as Bridgman put it in 1929 in *Harper's*. Leading physicists tried to restore that bottom, wrestling with the meaning of quantum mechanics in correspondence, articles, and textbooks. No consensus emerged. Bohr proposed complementarity, others invoked a statistical conception of reality. Schrödinger literally made waves while trying to turn quantum mechanics into a causal theory, while Einstein rebelled entirely, pinning his hopes on a full restoration of the Newtonian Moment. But no one shirked from addressing the fact that quantum mechanics raised big questions with philosophical implications.

Kaiser's tale begins right there. Hard-nosed scientists both in

Europe and the United States, he points out at the beginning of his book, insisted that their students understand the epistemologically bizarre aspects of quantum mechanics and "hone their own philosophical response." For a physicist of the 1930s, Kaiser told our class, having a philosophical position on quantum mechanics was part of what it meant to be educated.

The war changed everything. "The Second World War erased philosophy from American physics," Kaiser told our class. "Not from the American temperament, but from American physics. The incredible pressure of the Manhattan Project to build the atomic bomb, to engineer such a huge project, and then to engineer the huge scientific projects after that, forced the philosophy out of sight. It was left behind in the classrooms, too. Physics classes were once seminar-sized; now they were held in lecture halls before many times as many students, and the curriculum focused on solving problems. You have to teach differently in lecture situations than you do in seminars. Physicists learned to put their heads down, ignore philosophical tangents, and wring numbers from their equations as quickly as possible."

The lives of career physicists, he continued, morphed further during the Cold War, as they were recruited in increasing numbers to demanding work in industries and national labs. Rising enrollments continued to transform classroom physics, which now emphasized calculations and problem sets in textbooks and standardized exams. Kaiser called it "high-throughput pedagogy." Suddenly physics was a profession rather than a calling, and its practitioners were trained more like functionaries than scholars.

What, then, to say when asked how to interpret quantum mechanics? The answer was what, in 1955, Heisenberg dubbed the Copenhagen Interpretation. In a recent article, the Notre Dame science historian Don Howard has tracked the genesis of this phrase to the postwar period.[1] "Simply put," Howard writes, "the image of a unitary Copenhagen interpretation is a postwar myth, invented by Heisenberg." Heisenberg had been revered during the prewar years as one of the founders of quantum mechanics and played a strong role in its application to nuclear physics. However, as Howard writes, "[T]he person who was Bohr's favorite in the 1920s had become a moral exile from the Copenhagen inner circle in the postwar period, mainly because of the bitter rupture in Heisenberg's relationship with Bohr during his ill-fated visit to Copenhagen in September 1941 after taking over the leadership of the German atomic bomb project." A decade after the war's end, Heisenberg attempted to recover some of his prewar mantle by retelling the story of the philosophical discussions of the 1920s and 1930s in a way that papered over the differences between himself, Bohr, and others, and portrayed himself as a luminary. "What better way," Howard writes, "for a proud and once ambitious Heisenberg to reclaim membership in the Copenhagen family than by making himself the voice of the Copenhagen interpretation?"

The occasion was a volume of essays in Bohr's honor published in 1955. Heisenberg's contribution attributes a single interpretation to himself, Bohr, and others. "What was born in Copenhagen in 1927," Heisenberg writes, "was not only an unambiguous prescription for the interpretation of experiments, but also a language in which one spoke about Nature on the atomic scale, and in so far a part of philosophy."[2]

Once invented and named, says Howard, "the myth took hold as other authors put it to use in the furtherance of their own agendas." The Copenhagen Interpretation either became the convenient name for a supposed orthodox position and was assumed to be all but unassailable—or, for a few mavericks, the name of the position they were opposing. These mavericks included David Bohm, a proponent of hidden variables; Paul Feyerabend, a methodological anarchist; and Karl Popper, who championed objectivism against the supposed subjectivism of quantum mechanics. Howard could also have mentioned traditional Marxists, whose dialectical materialism opposed anything but a thoroughgoing objectivism; the official Soviet scientific line was anti–Copenhagen Interpretation.

Among physicists themselves, however, the Copenhagen Interpretation became a kind of placeholder name for whatever philosophical position would work to explain quantum mechanics, one that the physicists themselves need not investigate. Asking big questions was repressed, regarded as unnecessary and even handicapping. "The fundamental strangeness of quantum reality had been leeched out," Kaiser writes. The proper response to questions of interpretation was "Shut up and calculate." We've heard this same thought expressed in other ways, such as "Suck it up!" and "Deal!"

A philosopher knows that you can't get away with this. If you fail to address philosophical issues head-on they don't go away, they come round and bite you on the rear.

In mid-1975, after the physics bubble burst, funding plummeted, and jobs evaporated, a small coterie of California physicists with stymied careers rebelled against the no-big-questions, turn-the-crank approach to quantum mechanics and created the Fundamental Fysiks Group at Berkeley. Its aim, Kaiser writes, was "to recapture the sense of excitement, wonder, and mystery that had attracted them to physics in the first place, just as it had animated the founders of quantum mechanics."

They noted that the strange picture of the subatomic realm painted by quantum mechanics could not be linked up with the familiar everyday world in a satisfactory way. There just did not seem to be any suitable words or concepts that would clarify what was happening. The Fundamental Fysiks Group began drawing connections between concepts of the mystics and those that seemed to be involved in quantum physics. Several physicists joined in the conversation—not necessarily because these people were right, but because it seemed to be the only conversation around concerning the meaning of quantum mechanics.

One member of the group was Jack Sarfatti, a Brooklyn-born theoretical physicist who had received his PhD in 1969 from the University of California at Riverside, had taught for a while, but was no longer in academia. Another was Fred Alan Wolf, who had received his PhD from the University of California, Los Angeles, in 1963. The two loved to concoct imaginative if off-the-wall ways of explaining the difference between classical and quantum mechanics. One involved a creative interpretation of the Beatles' song "Being for the Benefit of Mr. Kite!" in which "Mr. K" (who "will challenge the world") stood for Boltzmann's constant (k, which is related to entropy), while "Mr. H" (who performs somersaults) stood for Planck's constant (h).[3]

John Clauser, a theoretical physicist who had received his PhD from Columbia University in 1969, also joined the group a few years after arriving at Lawrence Berkeley Laboratory. Clauser viewed the other group members as "a bunch of nuts, really," but was attracted because they provided "the only setting in which physicists could talk about the latest developments in quantum nonlocality."[4] Yet another group member was physicist Fritjof Capra, who received his PhD from the University of Vienna in 1966. While doing physics research at various places in California in the 1960s, Capra had heard radio lectures by Alan Watts, the foremost interpreter of Eastern philosophy to Western audiences at the time, then also living in California. "That's what I mean by cultural gears," Kaiser told our class. "Why was this movement born in that specific place at that specific time? Because its institutions, aside from including forefront physics laboratories, also included ones that promoted Eastern philosophies." Capra joined the Fundamental Fysiks Group after he arrived at the Lawrence Berkeley Laboratory in 1975.

The group was partly funded by a self-help guru who was born Jack Rosenberg but who renamed himself Werner Erhard—the first name after Werner Heisenberg. Erhard had recently become famous, and wealthy, for "Erhard Seminars Training" (or EST), one of the most successful personal growth initiatives at the beginning of the New Age movement. In addition to Erhard and EST, the group received support from Michael Murphy, cofounder of the Esalen Institute, a resort/retreat in Big Sur, California, known for its drop-dead views, fine food, hot tubs, and massages. Many of these Esalen workshops were aimed at New Age audiences, though they also attracted some physicists from Europe who, like members of the Fundamental Fysiks Group, had grown frustrated by their colleagues' avoidance of foundational topics in quantum theory. These included scientists like Bernard d'Espagnat and H. Dieter Zeh, whose visits to Esalen Kaiser mentions in his book. At the same time, Erhard contacted eminent physicists and offered to sponsor serious physics gatherings in his San Francisco mansion. Harvard physicist Sidney Coleman took Erhard up on it, and organized several Erhard-sponsored conferences on aspects of quantum mechanics in Erhard's residence in the late 1970s. One of us—Goldhaber—participated, and had the chance to chat for a few minutes with the self-help guru. Erhard was on his best behavior, because Coleman—who had denounced the fashion of linking quantum mechanics and Eastern philosophy as "the golden age of silliness"—had vociferously insisted that only serious physics be discussed. No silliness materialized, at least at that conference.

The quantum mysticism movement grew even after the Fundamental Fysiks Group disbanded in 1979. One landmark was the publication of Capra's book *The Tao of Physics*, published in 1975, which while initially rejected by twenty-five publishers went on to sell millions of copies. In it, Capra wrote that the connection between physics and Zen Buddhism is of critical significance, for it demonstrates the inadequacy of our present worldview and points to the need for a change so drastic as to amount to cultural revolution: "The survival of our whole civilization may depend on whether we can bring about such a change."[5] Capra's book made it into classrooms, as at last a book that could attract nonscience students to take physics.[6] A nonscientist, Gary Zukav, then produced *The Dancing Wu Li Masters*, which made the breezy announcement that "philosophically . . . the implications [of quantum mechanics] are psychedelic," and won an American Book Award.[7] Among the numerous other authors who espouse quantum mysticism is Alex Comfort (of *Joy of Sex* fame), in *Reality & Empathy: Physics, Mind & Science in the 21st Century*.[8] Quantum mysticism has appeared in movies; the baseball groupie played by Susan Sarandon in *Bull Durham* is a quantum mystic. One of the most over-the-top examples of quantum mysticism is the movie *What the Bleep Do We Know?* (2004), which cashes in on a wide variety of quantum fruitloopery.

Many of the Fundamental Fysiks Group's wackier concerns, such as connecting quantum physics with consciousness, did not pan out. The brain is "wet and warm," as scientists say, which kills the possibility of quantum effects. Still, Kaiser argues that the activities of the countercultural Fundamental Fysiks Group exerted an influence on mainstream science by motivating development of some of the weirder aspects of its theory and tests of some of its weirder experimental predictions. In an example of no-way physics, the group members thought that they could use quantum mechanics to devise a way to disprove Einstein's theory that faster-than-light communication was impossible. These efforts failed, Kaiser says, but had the positive outcome of leading them to the "no-cloning theorem," a proof that it is impossible to make an exact copy of a quantum state. This would become a key element of quantum cryptography. The Fundamental Fysiks Group also did much to rescue Bell's theorem from obscurity and thereby "planted the seeds that would eventually flower into today's field of quantum information science." Bell's theorem had been mentioned briefly in one physics textbook, but one not widely adopted at the time.[9] Although Clauser first experimentally tested Bell's theorem in 1972, before the group's formation in 1975, he was a charter member of the group, in frequent contact with its members even before they began meeting, and remained a key participant all the way through its lifetime.

The big story that Kaiser is telling, however, is how the interpretive issues resurfaced as big questions. In the California cultural atmosphere of the time, laden with discussions of Eastern philosophy, extrasensory perception, and psychic research, and enriched with hot tubs, drugs, and sex, this eclectic group of nonconformists (the "hippies" of the title) created a space outside the mainstream that relegitimized ("saved") the act of posing big questions. The climate in which they did so, however, also encouraged drawing connections between quantum mechanics and Eastern mysticism, which still makes physicists cringe.

Several physicists and philosophers attempted to counter quantum mysticism, but with explanations nearly as strange as what they were countering.[10] How then do we weed out what is fruitloopery from what is not? The answer requires analyzing the usage following clues provided by Kaiser.

Quantum mysticism was generated by the awareness that quantum mechanics forces us to reject a traditional understanding of objectivity. It seems to legitimize mystery, to celebrate incomprehensibility, which is how it often appears in popular culture. Consider an example from the novel *Brida*, by the Portuguese novelist Paulo Coelho. Overcoming her fears but still terrified, the eponymous protagonist has just explained to her boyfriend Lorens a recent mystical and magical experience that required a form of faith in the face of incomprehensibility that she calls the Dark Night. Shaking, she tells him she is sure he doesn't believe a word of her story. The boyfriend calmly puts two holes in a piece of paper, then fetches a cork from the kitchen. "Let's pretend that this cork is an electron, one of the small particles that make up the atom. Do you understand?"

Brida nods. "Right, well listen carefully," Lorens says. "If I had certain highly complicated bits of apparatus with me that would allow me to shoot an electron in the direction of that piece of paper, it would pass through the two holes at the same time, except that it would do so without splitting into two." The boyfriend is referring, of course, to the double-slit experiment discussed in Chapter 8. Brida says she doesn't believe it, for that's impossible. Lorens assures her that it is true. Brida then asks, "And what do scientists do when confronted by these mysteries?" Lorens's response:

> They enter the Dark Night, to use a term you taught me. We know that the mystery won't ever go away and so we learn to accept it, to live with it. I think the same thing happens in many situations in life. A mother bringing up a child must feel that she's plunging into the Dark Night, too. Or an immigrant who travels to a far-off country in search of work and money. They believe that their efforts will be rewarded and that one day they'll understand what happened along the way that, at the time, seemed so very frightening. It isn't explanations that carry us forward, it's our desire to go on.[11]

Setting aside the pseudoscience aspects of witchcraft in the background, the boyfriend's point is that if scientists can accept mystery—the Dark Night—so can we all. In this way, quantum mechanics, as one Stony Brook graduate student pointed out to us, fosters "a curious intersection of the Enlightenment scientific project and the Romantic insistence on ultimate mysteries."[12]

What's the Fuss?

It is at this point that a philosopher wonders what the fuss was all about. The "new epistemology" that quantum mechanics calls for: is it really an Eastern epistemology, or does the Western tradition have resources for explaining what is happening without celebrating the irrational and mystical? It certainly does. Quantum mechanics undermines a notion of objectivity based on nineteenth-century, Newtonian science—but only *that* notion. At the same time that quantum mechanics was emerging in the twentieth century, so was a notion of objectivity that was suitable for describing quantum objects. The term "objectivity" refers to an ideal of knowledge that is frequently characterized as a "view from nowhere," one that an observer might somehow achieve when standing completely apart and disconnected from what was being observed. Scientists' thoughts on how to approach this ideal in practice—as the historians of science Lorraine Daston and Peter Galison show in a book entitled simply *Objectivity*—evolved in both meaning and symbolism over time. But this ideal steered the ambition of science in the Newtonian Moment. Bohr challenged this ambition, at least in certain domains, in the Como talk in which he introduced the notion of complementarity. A "sharp separation" between quantum phenomena and the means of observation—the instrumental context in which you are observing the quantum phenomena—is not possible, he said.

Bohr was simply being a hard-headed empiricist. Every seeker of knowledge, whether in the ordinary or scientific realm, is tied to a particular place, utilizes particular means of perceiving, and is motivated by a particular set of questions. The means of observation—whether these be our ordinary senses, or instruments to extend our perception—are part and parcel of being able to have a world. These means allow us to experience objects we could not experience otherwise, and help to structure these objects. You cannot give a full account of these objects into which you are inquiring, that is, without talking about the process of inquiring itself. But the world is still experienced as being independent of human beings: we can't make it do what we want; the world pushes back; it may behave unexpectedly.

The new notion of objectivity was introduced at the beginning of the twentieth century by the German philosopher Edmund Husserl, founder of the philosophical movement known as phenomenology (though Hegel, writing almost a century earlier, had described some aspects of this notion). Phenomenology takes at its starting point that all inquiry—philosophical and scientific—is first and foremost a question of looking and discovering rather than assuming and deducing. In looking and discovering, an object is not seen from nowhere, but appears to someone—whether individual or community. Furthermore, the way the object is given, its particular appearance, is related to how it is looked at, as we saw in the Interlude to Chapter 8. The act of perceiving an object contains *anticipations* of other acts in which the same object will be experienced in other ways. That's what gives our experience of the world its depth and density. To perceive something as a presence in the world, as "objective," means to grasp it as never totally given, but as having an infinite number of appearances that are not grasped. Perhaps our original perception was misled, and our anticipations mere assumptions; still, we discover this only through looking and discovering—through exploring other appearances. In this view, objectivity does not mean standing outside an object and somehow grasping it from nowhere; it is rather being able to grasp the object so that one knows how it will be grasped from other standpoints. The objective point of view is not how no one would see some thing, or how one would see it from nowhere—but how everyone *could* in principle see it.

In science, perception of an object like an electron or planet involves instrumental mediation. When researchers propose the existence, say, of a new particle or asteroid, such a proposal involves a horizon of anticipations of that entity appearing in other ways in other circumstances, anticipations that can be confirmed or disconfirmed only by looking, in some suitably broad sense. For scientific objects, too, the horizon means that there is always more to come.

Scientists know this: it is obvious to them that phenomena appear differently when addressed by different instrumentation, and therefore that their conceptual grasp of the phenomena involves instrumental mediation and anticipation. Not only is there no "view from nowhere" of such phenomena, but there is also no Google Earth, or position from which we can tinker with the instruments to zoom in on every available profile, for there is no one privileged position, and the instrumentally mediated "positions" on which we are standing are changing. The fact that events may only take place in special laboratory contexts does not mean that they are abstract and unworldly. It's the other way around: the special laboratory contexts are what make these events part of the world, and therefore are responsible for their being of pressing concern to us.

How does this notion of objectivity explain puzzling features of quantum mechanics like wave-particle duality and superposed states? For one thing, phenomenology is not bothered by phenomena that are unlike conventional objects like tables and chairs; phenomenological description does not presuppose how something "is" but sets out to describe how it is. Our ordinary experience is full of phenomena that do not behave like ordinary things—emotions, the experience of nearness, the wind. Our ordinary experience even includes examples of plural, mixed, incomplete, or semiabstract realities. One is the famous duck-rabbit found in many philosophy and psychology textbooks. Here's a version:

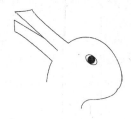

Set against the background of a pond, with rippling water waves and lily pads, it's unequivocally a duck. Set in a meadow, with grass and dandelions, it's equally surely a rabbit. In each, parts of the figure are meaningless or meaningful; for example, the bump to the right is a mouth in the pasture but just an indentation in the pond. The "state" is different. What "is" that figure "in itself"? Stripped of an environment, it has a kind of ambiguous—one might even say "blurred"—reality. If you demand to know what it really is, the answer is either absurd or ambivalent. You can't obtain a better answer by taking the image apart, considering pieces of it or the ink from which it is made. It's a whole.

The duck-rabbit is an analogue to the situation illustrated by Schrödinger's cat. The pre-box-opened cat's ψ-function superposes two possibilities—live cat and dead cat—in a way that is like the image apart from a background. Supplying a background—bringing it "into" the world—is like making an observation or taking a measurement. The figure morphing into an object that is unambiguous to human perception is like the "collapse" of the wave function. There is not much mystery in this process. The mystery is created only when one asks of the original figure, "What is it?" and demands the same kind of answer that one gets for a perceptual object. Philosophers know that certain kinds of questions are meaningless ("Can God create a stone that is heavier than He can lift?") because they anticipate an answer that cannot be given. Science is a prediction machine. When there's not enough information to make a prediction, all it can do is stop with what it has. Schrödinger's cat depicts those kinds of situations.

Let's take another example. A script or score has a funny kind of polymorphous reality. It, too, is incomplete and semiabstract, a description of potentially different kinds of incompatible events (different versions of *Hamlet* or *Julius Caesar*, say) and a tool for producing more such events. To turn a script or score into a real, worldly event, one has to add the context and produce the event, which means making a set of decisions about places, performers, and props. The outcome of one set of decisions—playing

Hamlet as mad, say—may look utterly different from the outcome of another set of decisions—making him rational and manipulative. Lines of dialogue that are key in one performance are less meaningful in another. But this does not mean that the play producers and directors "create" Hamlet from scratch. Far from it, they are tightly constrained. A script or score has a certain kind of semiabstract reality because it both limits us and lets us play with it in new ways. But it's not the kind of reality that a specific performance has.

The ψ-function of Schrödinger's equation is another kind of semiabstract object that can be realized in different ways depending on the environment; it may, but will not necessarily, involve measurement situations in which human purposes and decisions come into play. Schrödinger's gag set up a quantum answer to a classical question. It envisioned a cat with a ψ-function, and asked if it were real. That's like asking whether a duck-rabbit is a duck or a rabbit.

But we know there cannot be such a ψ-function. A cat is always interacting with the world. The box it's in is full of things that observe or "measure" the cat, providing information of its existence. The ratio of oxygen to carbon dioxide molecules, for instance, measures how many breaths it's taken. Every minute it is living the heat of its body radiates into the environment. The box is full of what amount to cat detectors.

Albert Einstein once asked Abraham Pais if he believed that the moon existed only when looked at.[13] But of course the world is full of moon detectors that are not humans: tides, satellites, moonlight, shadows, and so forth. As the philosopher Marshall Spector put it:

> Does the moon exist at a time for which *no objective physical events whatsoever*—photographic images, tides, or anything of the kind—can be interpreted as allowing us to conclude the presence of this allegedly massive, reflective object relatively near

the Earth? No, under such circumstances it does not! And what a shock this would indeed be! But note that this is rather a far cry from the *apparent* claim that if we *humans* all *chose* to avert our *eyes* the moon would go out of existence, and is recreated by us (*each* of us) at the *instant* we choose to look up again. Mind has not created matter.[14]

The world is implicated in a dense causal network. We are entitled to say that things in that network exist, but the same is not the case for things outside that network. The astounding thing about quantum mechanics is that it entitles us to say *some* things about things outside that network, that is, that they can be realized in this way or that way, but those two ways have a certain relation to each other. Certain potentialities have a particular range of actualities.

The duck-rabbit is like probability waves and other features of quantum phenomena in that it has a shape, and is real, but in a different way from ducks and rabbits. The same is true of quantum phenomena.[15] The new notion of objectivity is not a position for no one, but a position for everyone. The objective is something that has to make sense, in principle at least, to everyone else doing research.

The Now Moment

> Never in the history of science has there been a theory which has had such a profound impact on human thinking as quantum mechanics; nor has there been a theory which scored such spectacular successes in the prediction of such an enormous variety of phenomena.
> —Max Jammer, *The Philosophy of Quantum Mechanics*

Shortly before finishing this book, we gave a talk at Stony Brook University's Humanities Institute, where we encountered a disgruntled questioner who expressed disbelief that science could contribute anything to culture. What, he asked indignantly, did quantum mechanics really add to the cultural conversation? Maybe it's given us fancy new metaphors for old issues; but if these are stripped away what's really lost? Wouldn't artists and philosophers find other ways to express themselves? Scientists need the uncertainty principle. Do poets?

The most direct way to respond is to propose a thought experiment: Imagine that the Newtonian Moment never ended. Suppose that a few years after Planck proposed the idea of the quantum, he abandoned it after he and others managed to explain much of what scientists were discovering without it. Let's say they developed a reasonably satisfactory way to explain atoms as consisting of little bits of matter that connect to each other via tiny springs—Hooke's Law springs, we called them in Chapter 1—whose elasticity is a function of the applied force (we love the irony of having the Newtonian Moment rescued by an invention concocted by Newton's arch-nemesis Hooke). Let's imagine that the theory of Hooke's Law springs, plus other Newtonian principles and thermodynamics, somehow proves successful enough to confirm Planck's initial belief that the quantum was merely a mathematical trick, and soon it disappears from science, remembered only by historians. Quantum theory, that is, meets a similar fate as phlogiston theory, the long-discarded speculation that a substance called phlogiston causes fire. In our scenario, scientists decide that the microworld obeys the same basic laws as the macroworld. Classical differential equations remain every physicist's principal calculational tool.

In the early twentieth century, physicists therefore experience no particular sense of anxiety or crisis. A few mysteries remain, especially regarding the nature of light, and why the elementary bits of matter of the same type are all identical, but physicists assume that eventually satisfactory answers will emerge. Their field continues to be orderly, predictable, and unified, its laws and properties the same throughout all areas and scales. They belong to a growing, congenial community, and remain confident of eventually being able to "take themselves out" of measurements to reach an objective description of nature and each of its properties. At the beginning of the second decade of the twentieth century, Ernest Solvay becomes famous for sponsoring a series of conferences on international business methods and practices rather than physics.

Because there is no first quantum revolution there is no second one either, and physicists spend the years 1925 to 1927 mainly fretting about the origin and nature of cosmic rays. Popular culture

never learns about quantum leaps, wave-particle duality, the uncertainty principle, complementarity, superposition, Schrödinger's cat, spooky action at a distance, or parallel worlds. There's no double-slit experiment and no Bell's theorem. The Marquis de Laplace is hailed as a superstar. Einstein is celebrated for his courageous stand defending causality against a coterie of shortsighted physicists with small imaginations—Bohr, Heisenberg, and Dirac among them—who were momentarily seduced into abandoning it, who faced mockery from colleagues for doing so for the rest of their careers.

In the 1960s, plummeting enrollments in physics, its declining reputation thanks to its role in defense research, as well as what many regarded as the profession's cynical tilt toward applied science, led a small group of disaffected California physicists to explore the possibility that the practice of physics at the nano scale could reveal deep and mystical truths. A group calling itself the Infynytzimal Fysiks Fraternity begins meeting at Berkeley. Physics, its members pointed out, had never managed to explain successfully why any particular atom was identical to any other, and they were fond of paraphrasing Alfred, Lord Tennyson:

> Little atom—but if I could understand
> What you are, electrons and all, and all in all,
> I should know what God and man is.

Other members plumbed the microworld for parallels with the truths of Eastern religions. Hadn't the sixty-century BC. Chinese philosopher Lao Tsu, the founder of Taoism, predicted nano physics when he declared that "All difficult things have their origin in that which is easy, and great things in that which is small"? One member of the Infynytzimal Fysiks Fraternity writes a book called *The Dancing Nano Ninjas.* Scientists criticize it for its over-the-top prose and loose imagery, but it acquaints much of the public with physics research, sells a million copies, and wins a National Book Award.

So how does this counterfactual world stack up against ours? Not very well.

From a purely scientific point of view, physicists still would find much of the world—its blueprint—impossible to understand. Light must remain a mystery, as well as the absolute identity of like bits of matter. Another mystery is matter's stability; if things are built out of Hooke-like springs, why are they solid and not jiggly? Uncertainty, as we have seen, is a resource rather than a bug in quantum theory, because things that are squishy in classical physics can be hard in quantum mechanics in ways that completely explain matter's rigidity. Quantum mechanics, in short, explains the foundations of the world. Many questions of greater scope also remain mysterious, including life and its development. Quantum mechanics explains two essential features of the DNA molecule: both its extreme stability and its occasional ability to mutate. It describes the DNA molecule as robust in actual living environments, for it explains why it takes a large chunk of energy to rearrange some portion of the molecule. Yet quantum mechanics also guarantees that, every once in a while, the finite amount of energy needed to change the molecule will occur—and then that that change or mutation will remain stable for some time. The resulting variability from one member of a species to another generates the resourcefulness of life in responding to changes in external circumstances. The fact that the DNA molecule is an intrinsically quantum object is therefore crucial to its success. No Newtonian mechanism could explain this.

From a technological point of view, too, this counterfactual world would be greatly impoverished compared to ours. Quantum mechanics has vastly expanded our ability to manipulate the world, because knowing its blueprint allows us to make new kinds of things. We have seen that there have been two quantum scientific revolutions—but there have been as well two quantum technological revolutions. In the first, quantum theory found practical uses through its relevance to solid state physics, vastly expanding the ability to construct and manipulate materials, to fashion things like transistors, semiconductors, and other materials essential to mobile phones, iPads, and other devices of modern information technology.[1] Quantum mechanics plus massive computational ability has made it possible to connect quantum microstructures and Newtonian macrostructures, and the design and discovery of so-called quantum materials and quantum devices continues to be a growing field. This first quantum technological revolution exploited the "nonweird" aspects of quantum mechanics. The second quantum technological revolution, just now beginning, exploits the "weird" aspects of quantum mechanics, such as superposition and entanglement, in devices such as quantum computers and cryptography. These may be on the way to becoming commercially viable.

Our adversary at the Humanities Institute, however, was not challenging the fact that the quantum has given us a better grip on the material world, but asking how it could possibly give a better grip on ourselves. Wouldn't the humanities fare just as well in this counterfactual world?

No, we answer. It is true that, to some extent, the quantum's impact on artists, writers, and philosophers was that it helped them free themselves from their own Newtonian-inspired misconceptions. To this extent quantum mechanics has made "negative philosophical discoveries," by showing that some of what we thought were philosophical truths were just plain empirically wrong. Reality may smell different in the quantum world, as Einstein said, but the smell is ultimately fresher. Quantum mechanics, for instance, has helped rid philosophy of the specter of a Laplacean ideal of knowledge, an "intelligence sufficiently vast" that it could see and describe things as if from no particular time and place, and rid philosophy as well of the vision of a unified science, of a too-narrow conception of phenomena, and of an impossible objectivity. But, like Newtonian mechanics, quantum mechanics has helped us attend to the world, and ourselves, in a refreshing new way. As we have shown throughout this book, it has provided new images to focus on ancient human yearnings and issues, and given us a terminology and language that helps guide us what to look out for, and what it is all right to overlook. Our world, unlike that of our ancestors, is full of what Updike called "gaps, inconsistencies, warps, and bubbles." Unlike our ancestors, we have to cope with what Dawkins called "the queerness of the very small, the very large and the very fast," in ways that we cannot digest with Newtonian sensibilities. As a result, our novelists and poets have to address a reality whose surface is less like the smooth geometry of the Newtonian world and more like that of a kettle of boiling water. The world, and ourselves, are both stranger than we recognized—and we were helped to this recognition in a strange way, with the assist of a set of imagery and language that sprang from a remote corner of physics over a century ago. We have traced how many of these images passed outside physics into other domains and even to ordinary language.[2] In some cases this passage was driven by metaphorical compulsion, arising from a discontent that drives us to put words or concepts to our experience, the sense that the existing and conventional words and concepts don't fit—that there is more to be said—and the search for other words and concepts that might. In other cases the passage was exploitative, capitalizing on the cultural authority of science, its connotations of depth and fundamentality. There is no substitute for being able to have the *right* image, one that best captures our intuitions. Quantum mechanics has helped rekindle our awe at a cosmos that, as we quoted the *New York Times* as saying in Chapter 2, is not a machine that moves in a predictable way, but one whose phenomena can be more whimsical than fairies and more wonderful than Aladdin's lamp, and whose mystery and beauty await a new Lucretius whose inspiration comes from water wells fed by the like of Planck, Einstein, Schrödinger, and Heisenberg. In helping to sensitize us to the warps and bubbles of the world, it has helped create the prospect of what Updike suggested was a "new humanism."

In Chapter 9 we cited Heisenberg's remark that "Almost every progress in science has been paid for by a sacrifice, for almost every new intellectual achievement previous positions and conceptions had to be given up. Thus, in a way, the increase of knowledge and insight diminishes continually the scientist's claim on 'understanding' nature." We disagree. Scientists and engineers—and everyone else—have given up very little, and apply Newtonian intuitions successfully every day. The advent of quantum mechanics has meant only that we've had to develop new intuitions for special domains; what we've had to sacrifice is only the dream of a single, comprehensive set of intuitions that works in all scales and domains. The Newtonian and Quantum Moments belong together and could not exist without each other. Their interaction is productive rather than antagonistic and exclusive.

In many respects, indeed, the Newtonian Moment and the Quantum Moment are more similar than they might appear. The Newtonian world knows uncertainty and unpredictability; the quantum world is strongly shaped by a highly precise mathematical determinism that accounts for the world's rigidity. That's why scientists could get by without the latter until well into the twentieth century. Yet the characters of the two worlds are vastly different. The Newtonian world is like a rigid building that floats without a foundation; the quantum world is like a firmly built treehouse, connected to a ground, but able to sway in the breeze. Still more significantly, the two worlds provide human beings with different answers to the three key philosophical questions: What can we know? How we should act? What might we hope for?

Understanding and appreciating quantum language and imagery—along with the ability to recognize its misuse in fruitloopery—is part of what it means to be an educated person today. Such an education involves learning elements of both science and the humanities. Acquiring that kind of education requires crossing many traditional disciplinary boundaries, but such is the entangled state of literacy in the modern world. Facing the Quantum

Moment requires a new framework for the humanities of the twenty-first century.

In the first chapter, we quoted the remark by historians Betty Dobbs and Margaret Jacob that the Newtonian Moment provided "the material and mental universe—industrial and scientific—in which most Westerners and some non-Westerners now live, one aptly described as modernity."[3] Yet this universe is slowly changing. What is the right way to describe what comes after it? We loathe the commonly used term "postmodernity," which is bandied about by many different people assigning it many meanings. We also dislike the way the term is embraced by many who, as exposed by Sokal in his notorious hoax we mentioned in Chapter 8, are scientifically illiterate or even antiscientific—for it is clear that science is intimately entwined in the framework of this post-Newtonian cultural universe, a fact that must be faced by anyone hoping to come to grips with it. Is it possible, we ask our class, that the right term for describing what comes after the Newtonian Moment is the Quantum Moment?

Notes

Introduction

1. David Javerbaum, "A Quantum Theory of Mitt Romney," *The New York Times*, March 31, 2012, p. SR4.
2. Alison Bechdel, *Fun Home: A Family Tragicomic* (New York: Houghton Mifflin, 2006), p. 104.

Chapter One: The Newtonian Moment

1. John Green and David Levithan, *Will Grayson, Will Grayson* (New York: Dutton, 2010), pp. 82–83.
2. Ryan Boudinot, *Blueprints of the Afterlife* (New York: Grove Press, 2012).
3. Mordechai Feingold, *The Newtonian Moment: Isaac Newton and the Making of Modern Culture* (New York: Oxford University Press, 2004), p. xi.
4. See for instance Loren R. Graham, "The Socio-Political Roots of Boris Hessen: Soviet Marxism and the History of Science," *Social Studies of Science* 15, no. 4 (November 1985), pp. 705–22.
5. Feingold, *The Newtonian Moment*, p. xiii.
6. Revolutionary scientific theories often have this effect; for an account of the cultural impact of Darwin's evolutionary theory see Monte Reel, *Between Man and Beast: An Unlikely Explorer, the Evolution Debates, and the African Adventure That Took the Victorian World by Storm* (New York: Doubleday, 2013).
7. In *The Machiavellian Moment: Florentine Political Thought and the Atlantic Republican Tradition* (Princeton: Princeton University Press, 1975), for instance, the historian J.G.A. Pocock details how the writings of Niccolò Machiavelli (1469–1527) were a response to the baffling and seemingly insoluble fact that virtuous republics were always threatened with destruction, much of it apparently self-caused—and how by explicitly

confronting and accounting for this fact, Machiavelli's writings opened a new landscape for political thought and action, one that long outlived Machiavelli himself. To appreciate the Machiavellian Moment, Pocock demonstrates, we have to understand not only Machiavelli's writings, but also the historical situation to which he responded, and how he transformed it.

8. Pierre Laplace, *A Philosophical Essay on Probabilities*, tr. F. Truscott and F. Emory (New York: Dover, 1951), p. 4.

9. William Thomson (Lord Kelvin), *Kelvin's Baltimore Lectures and Modern Theoretical Physics: Historical and Philosophical Perspectives* (Studies from the Johns Hopkins Center for the History & Philosophy of Science), ed. R. H. Kargon and P. Achinstein (Cambridge: MIT Press, 1987), p. 206.

10. Betty Jo Teeter Dobbs and Margaret C. Jacob, *Newton and the Culture of Newtonianism* (Atlantic Highlands, NJ: Humanities Press, 1985), p. 123.

11. Newton's life and work have been carefully and extensively documented and analyzed by science historians. Sources we found useful for classroom purposes were Dobbs and Jacob, *Newton*; and I. Bernard Cohen, *The Birth of a New Physics* (New York: Norton, 1985).

12. The classic example is a book by the British essayist Walter Bagehot, *Physics and Politics* (1872), which explored and extended the implications of Newtonian ideas for political conceptions, incorporating such post-Newtonian ideas as conservation of mass and energy and atomic behavior.

13. Richard Rorty, *Objectivity, Relativism, and Truth* (Cambridge: Cambridge University Press, 1991), p. 22.

14. Small wonder that Newtonian cults developed that were akin to secular religions, among both scholars and ordinary folk. One example was the freemasons, who established an alternative cultural experience of the day that emulated the order and rationality of the Newtonian universe. In Masonic lodges, Newton's system came to be "worshipped by literate elites who, regardless of how little science they actually knew and if only for an evening, turned themselves into priests officiating at ceremonies invented to honor the Grand Architect of the Universe and His magnificent creation" (Dobbs and Jacob, *Newton*, p. 104).

15. See for instance A. S. Barnhart, "The Exploitation of Gestalt Principles by Magicians," *Perception* 39 (2010), pp. 1286–89.

16. Richard Dawkins, Why the Universe Seems So Strange," http://www.ted.com/talks/richard_dawkins_on_our_queer_universe.html, accessed September 9, 2013.

17. Rudolf Carnap, Hans Hahn, and Otto Neurath, "The Scientific Conception of the World: The Vienna Circle," in *Philosophy of Technology: The Technological Condition*, ed. R. Scharff and V. Dusek (Oxford: Blackwell Publishing, 2003), p. 89.
18. See for instance Stephen Brush, "Irreversibility and Indeterminism: Fourier to Heisenberg," *Journal of the History of Ideas* 37 (1976), pp. 603–30.
19. Voltaire to Pierre-Louis Moreau de Maupertuis, October 1732, in *Voltaire's Correspondence*, vol. 2, ed. T. Besterman (Geneva: Voltaire Institute and Museum, 1953), p. 382.
20. As in Chapter 7 of Cohen, *The Birth of a New Physics*.
21. Brush, "Irreversibility."
22. Quoted in Brush, "Irreversibility," p. 611.

Chapter Two: A Pixelated World

1. John Updike, "Notes and Comment," *The New Yorker*, December 9, 1967, p. 51.
2. Planck's struggles to protect science during the Nazi regime are outlined in Philip Ball, *Serving the Reich: The Struggle for the Soul of Physics under Hitler* (London: Random House, 2013).
3. J. L. Heilbron, *The Dilemmas of an Upright Man: Max Planck as Spokesman for German Science* (Berkeley: University of California Press, 1986), p. 3. Other biographies of Planck and his work: H. Kragh, *Quantum Generations: A History of Physics in the Twentieth Century* (Princeton: Princeton University Press, 1999); and T. S. Kuhn, *Black-Body Theory and the Quantum Discontinuity: 1894–1912* (Oxford: Clarendon Press, 1978).
4. Max Planck, *Scientific Autobiography and Other Papers* (London: Williams & Norgate, 1950), pp. 34–35.
5. Albert Einstein to C. Habicht, May 1905, in *The Collected Papers of Albert Einstein*, vol. 5, tr. A. Beck (Princeton: Princeton University Press, 1995), Document 28, p. 21.
6. The First Solvay Conference, as it was called, was attended by 21 eminent European physicists: 7 from Germany (including Nernst, Planck, Wien, and Rubens), 2 from England (including Jeans of the Rayleigh-Jeans law), 6 from France (including Henri Poincaré and Madame Curie, the only woman), 2 from Austria (including Einstein, then in Prague, then part of the Austro-Hungarian empire), 1 from Belgium, 2 from the Netherlands, and 1 from Denmark.

7. Quoted in Max Jammer, *The Conceptual Development of Quantum Mechanics* (New York: McGraw-Hill, 1966), p. 59.

8. Quoted in Guido Bacciagaluppi and Anthony Valentini, *Quantum Theory at the Crossroads: Reconsidering the 1927 Solvay Conference* (Cambridge: Cambridge University Press, 2009, p. 4).

9. Quoted in Jammer, *Conceptual Development*, p. 54. This, remarked Jammer, may well be "one of the earliest, if not the earliest, renunciations of the universal validity of classical mechanics," which alone would make the conference a seminal moment in the history of science.

10. Henri Poincaré, "The Quantum Theory," in *Mathematics and Science: Last Essays*, tr. J. W. Bolduc (New York: Dover, 1963), p. 88. See also Russell McCormmach, "Henri Poincaré and the Quantum Theory," *ISIS* 58 (1967), pp. 37–55.

11. See for instance Maila L. Walter, *Science and Cultural Crisis: An Intellectual Biography of Percy Williams Bridgman (1882–1961)* (Redwood City, CA: Stanford University Press, 1991), p. 81.

12. Katherine Sopka, *Quantum Physics in America: 1920–1935* (New York: American Institute of Physics and Tomash Publishers, 1976), p. 39.

13. Werner Heisenberg, *Physics and Beyond: Encounters and Conversations* (New York: Harper & Row, 1971), p. 26.

14. Quoted in Sopka, *Quantum Physics in America*, p. 35; evidently it is the beginning of a student presentation found in Kemble's papers.

15. Robert Millikan, "Einstein's Photoelectric Equation and Contact Electromotive Force," *Physical Review* 7 (1916), p. 18. See also G. Holton, "R. A. Millikan's Struggle with the Meaning of Planck's Constant," *Physics in Perspective* 1, no. 3 (1999), pp. 231–37.

16. Paul Epstein, *Physikalische Zeitschrift* 17 (1916), p. 148; *Annalen der Physik* 50 (1916), p. 489.

17. Paul Heyl, *New Frontiers of Physics* (New York: D. Appleton, 1930), p. 67.

18. Jammer, *Conceptual Development*, pp. 36–37.

19. On April 3, 1916, for instance, Professor George Pegram of Columbia University gave a talk at the American Museum of American History in New York City entitled "Are There Atoms of Light? The Quantum Theory." The number and audiences for such talks grew as more news of the quantum reached the public.

20. *Manchester Guardian*, January 23, 1929, p. 9.

21. Consider, for instance, the following dialogue from Mark Leyner's 1992 novel *Et Tu, Babe*, in which a certain General Lopez is being interrogated by unidentified interrogators:

–General, one final question. Do you have any tattoos?

–Yes, sir.

–On what part of your body and of what? .

–I have E = nhf (Max Planck's formula for the energy in radiation) tattooed on my penile glans.

22. "Science Needs the Poet," *The New York Times*, December 21, 1930, p. 47.

23. Paul Hartman, *A Memoir on The Physical Review: A History of the First Hundred Years* (Woodbury, NY: AIP Press, 1994), pp. 43–44.

24. Max Planck, "On an Improvement of Wien's Equation for the Spectrum," *Annalen der Physik* 1 (1900), p. 730.

25. Max Planck, *Scientific Autobiography and Other Papers* (New York: Philosophical Library, 1949), p. 41.

26. Max Planck, "The Origin and Development of the Quantum Theory," Nobel Prize Lecture, 1922.

27. Max Planck to R. W. Wood, October 7, 1931, reproduced in A. Hermann, *The Genesis of Quantum Theory (1899–1913)*, tr. C. Nash (Cambridge: MIT Press, 1971), p. 23.

28. Jammer, *Conceptual Development*, p. 18.

29. Max Planck, "On the Theory of the Energy Distribution Law of the Normal Spectrum," *Verhalungen der Physikalischen Gesellschaft* 2 (1900), p. 202.

30. Jammer, *Conceptual Development*, p. 22.

31. So when was the quantum born? Discoveries sometimes do not emerge fully recognized for what they are, and the quantum is an illustration. Some people peg its birth to the October 19 formula. Jammer's date is May 1898, when Planck made his first calculations that, when carefully considered, require the quantum. Other historians, including Kragh and Kuhn, prefer a December 1900 birthday or even later. The latest date would be Einstein's 1905 paper, for Einstein recognized that the quantum applies to the entire field—which is why it can "knock" electrons out of materials.

32. Albert Einstein, "On a Heuristic Viewpoint Concerning the Production and Transformation of Light," *Annalen der Physik* 17 (1905), pp. 132–48, reproduced in *The Collected Papers of Albert Einstein*, vol. 2, tr. A. Beck (Princeton: Princeton University Press, 1989), Document 14, pp. 86–103. This paper was indeed revolutionary. But it was also conservative, in the sense that Einstein, like Planck, had kept as much of the old theory as possible, explaining things by incorporating as few add-ons as he could get away with. Planck had kept

classical mechanics and thermodynamics intact, and consistent with blackbody experiments, by adding the quantum idea to his resonators; Einstein kept classical particles and classical waves intact, and consistent with photoelectric data, by proposing to extend the quantum idea just a bit further, to the photons themselves. A revolution was brewing, but its creators were inching it forward with the tiniest moves they could.

Chapter Three: Quantum Leaps

1. George Gamow, *Thirty Years That Shook Physics* (New York: Doubleday, 1966), p. 56.
2. Bohr, whose PhD thesis was an attempt to understand the electrical properties of matter by picturing the mobile electric charges as a classical electron gas, was convinced that some kind of quantum description was necessary at the atomic scale. He set out to find such a picture, and showed that for the simplest atom, hydrogen, the spectrum of light emitted when the atom is excited could be explained by assuming that the electron goes in orbits that have discrete energies, with nothing in between allowed.
3. John L. Heilbron, "The Scattering of Alpha and Beta Particles and Rutherford's Atom," *Archive for History of Exact Sciences* 4 (1968), p. 304.
4. See Heilbron, "The Scattering."
5. Quoted in Abraham Pais, *Niels Bohr's Times, In Physics, Philosophy, and Polity* (New York: Clarendon Press, 1991), p. 144.
6. James Jeans, *Report on Radiation and the Quantum-Theory* (London: The Electrician Publishing Company, 1914), pp. 79–80.
7. Cited in Jammer, *Conceptual Development*, p. 77.
8. Niels Bohr, "On the Constitution of Atoms and Molecules, Part I," *Philosophical Magazine* 26 (1913), pp. 1–24; "On the Constitution of Atoms and Molecules, Part II: Systems Containing Only a Single Nucleus," *Philosophical Magazine* 26 (1913), pp. 476–502; "On the Constitution of Atoms and Molecules, Part III: Systems Containing Several Nuclei," *Philosophical Magazine* 26 (1913), pp. 857–75.
9. Pais, *Niels Bohr's Times*, p. 147.
10. Jeans, *Report*, p. 1.
11. Earlier suggestions of jumps by Max Planck (*Ann. Phys.* 342 (4) 642 (1912)), and Henri Poincaré (see Mathematics and Science: Last Essays, Dover 1963) apparently were unknown to Bohr.
12. These include Bertrand Russell's book, *The ABC of Atoms*, in 1923, and James Jeans's Rouse Ball lecture of 1925.

13. Waldemar Kaempffert, *The New York Times,* January 11, 1931.
14. "Down the Spillway," n. a., *The Sun,* July 4, 1929, p. 6.
15. "At Random," *The Observer,* November 10, 1929, p. 15.

Chapter Four: Randomness

1. BBC News, December 4, 2002, http://news.bbc.co.uk/2/hi/uk_news/ 2541761.stm, accessed September 19, 2013.
2. David Lindley, *Uncertainty: Einstein, Heisenberg, Bohr, and the Struggle for the Soul of Science* (New York: Doubleday, 2007), p. 30.
3. "[M]olecular science," wrote Maxwell in 1873, "teaches us that our experiments can never give us anything more than statistical information, and that no law deduced from them can pretend to absolute precision" ("Molecules," *Nature* [September 1873], pp. 437–41). Historians have indeed traced Maxwell's comfort with the idea of relying on statistics in natural science to the growing use of statistics in social science. See Theodore M. Porter, "A Statistical Survey of Gases: Maxwell's Social Physics," *Historical Studies in the Physical Sciences* 12, no. 1 (1981), pp. 77–116, at p. 79.
4. In the late nineteenth century, Maxwell, Boltzmann, and other scientists struggled to find principles to connect small-scale, reversible, Newtonian behaviors with large-scale, irreversible behaviors (such as those treated by thermodynamics). Boltzmann, for instance, introduced a constant that now bears his name—represented as k—that relates the temperature T of a substance to the average energy of an atom of that substance at the same temperature. But what these scientists found only served to emphasize the differences between microworld and macroworld. On small scales, things are reversible and predictable—but on large scales, statistics rule. Newton's laws + objects made of myriads of pieces + laws of probability = the arrow of time and an increase in entropy (disorder). The result was to create what Brush called a "kinetic worldview" among thermodynamic scientists, whose spirit was much different from that of traditional Newtonian physics. Thermodynamics also produced a problem called the "Gibbs paradox," discussed in the next chapter.
5. Einstein's idea was that light is always localized in packages—later called photons—like streams of billiard balls rather than of water. Precisely because these photons are localized, when they strike atoms they can knock out electrons—this would have been hard to understand on the basis of Planck's original idea—and you can calculate the maximum

energy that one of these electrons can have by using Planck's formula for the energy of an incident photon colliding with an atom. The electron loses some energy getting out of the material, so the formula includes the maximum amount of energy the electron can have—Planck's hv— minus the "work function" representing the energy it takes for the electron to "climb out" of the material. But nothing in Einstein's formula told you the probability that a photon will actually knock out an electron.

6. A. Einstein to M. Besso, August 11, 1916, in *The Collected Papers of Albert Einstein*, vol. 8, tr. A. Hentschel (Princeton: Princeton University Press, 1998), Document 250, p. 243.

7. A. Einstein to M. Besso, September 6, 1916, in *Collected Papers*, vol. 8, Document 254, p. 246.

8. Once again, Planck's paper of 1912 (mentioned earlier) introduced probability, but Einstein did not refer to Planck, even though by then they were neighbors in the same university, and it was Einstein's work that influenced what came later.

9. "The 'Thirties,'" *The New York Times*, December 29, 1929, p. E4.

10. Cathryn Carson, Alexei Kojevnikov, and Helmuth Trischler, "The Forman Thesis: 40 Years After," in *Weimar Culture and Quantum Mechanics: Selected Papers by Paul Forman and Contemporary Perspectives on the Forman Thesis*, ed. C. Carson, A. Kojevnikov, and H. Trischler (London: Imperial College Press, 2011), p. 3.

11. Paul Forman, PhD dissertation, University of California, Berkeley, 1967.

12. "Quite apart from its wealth of information on the great inflation, the consequences of the postwar isolation of Germany, intellectual life as a substitute for political and military power, anti-Semitism in the universities and so on," Heilbron wrote, "it demonstrated that, and also how, the powerful analytical tools and bibliographical resourcefulness of general historians, and also their results, could be brought to bear on the history of science.... Forman revealed to me that the history of science is history and that the writing of history was as demanding, creative and rewarding a discipline as the doing of science" (J. L. Heilbron, "Cold War Culture, History of Science and Postmodernity: Engagement of an Intellectual in a Hostile Academic Environment," in *Weimar Culture and Quantum Mechanics*, p. 11).

13. *Archive for History of Exact Sciences* 6 (1969), pp. 38–71.

14. *Historical Studies in the Physical Sciences* 3 (1971), pp. 1–115, reproduced in *Weimar Culture and Quantum Mechanics*, pp. 87–201, from which all quotes are taken.

15. Carson, Kojevnikov, and Trischler, eds., *Weimar Culture and Quantum Mechanics*, p. 2.

16. The most frequently cited of the three papers is "Quantentheorie des Strahlung," *Physikalische Zeitschrift* 18 (1917), pp. 121–28, reprinted as "On the Quantum Theory of Radiation," in *Sources of Quantum Mechanics*, ed. B. L. van der Waerden (Amsterdam: North Holland Publishing Co., 1967), pp. 63–78. This is a reprint of a paper in the *Physik. Gesell. Zurich Mitt.* 16 (1916), pp. 47–62; the third paper is "Strahlungsemission und Absorption nach der Quantentheorie," in *Verh. Deutsch. Phys. Ges.* 18 (1916), pp. 318–23.

17. Einstein, *The Collected Papers of Albert Einstein*, vol. 6, tr. A. Engel (Princeton: Princeton University Press, 1997), Document 34, p. 216.

18. Albert Einstein to Michele Besso, March 9, 1917, in *Collected Papers*, vol. 8, Document 306, p. 293.

19. Albert Einstein to Max Born, January 27, 1920, in *The Collected Papers of Albert Einstein*, vol. 9, tr. A. Hentschel (Princeton: Princeton University Press, 2004), Document 284, p. 237.

20. As Jammer points out on p. 113 of *Conceptual Development*, however, Einstein viewed this as a highly preliminary conclusion.

21. Albert Einstein, "Physics, Philosophy, and Scientific Progress," *Physics Today*, Vol. 58, June 2005, p. 46.

Chapter Five: The Matter of Identity

1. A famous story about the mathematician-philosopher Bertrand Russell has him lecturing to a general audience about logic, and stating that if the postulates of a mathematical system are logically contradictory, it is possible to prove any conclusion at all. A skeptic in the audience shouted out, "One equals two: prove that you are the Pope!" Unfazed, Russell replied, "I and the Pope are two; therefore I and the Pope are one." Russell was taking advantage of the equivalence between equality and identity: "I and the Pope are one" represents identity, whereas "I am the Pope" is the more literal way to represent equality.

2. Peter Pesic, *Seeing Double: Shared Identities in Physics, Philosophy, and Literature* (Cambridge: MIT Press, 2002), p. 98.

3. Pesic, *Seeing Double*, p. 14.

4. Robert P. Crease and Charles C. Mann, *The Second Creation: Makers of the Revolution in 20th Century Physics* (New York: Macmillan, 1986), p. 95.

5. Wolfgang Pauli, *Zeitschrift für Physik* 31 (1925), p. 765.

6. Pauli also assumed that any single-particle state, where the label of the

state included one extra two-valued label, today recognized as electron spin, quite analogous to the photon polarization, could be occupied by at most one electron.

7. See John L. Heilbron, "The Origins of the Exclusion Principle," *Historical Studies in the Physical Sciences* 13 (1983), p. 261.

8. Abraham Pais, *Inward Bound: Of Matter and Forces in the Physical World* (New York: Oxford University Press, 1986), p. 272.

9. Wolfgang Pauli, *Zeitschrift für Physik* 31 (1925), pp. 765–85.

10. Henry Margenau, "The Exclusion Principle and Its Philosophical Importance," *Philosophy of Science* 11 (1944), pp. 187–208.

11. P.A.M. Dirac, *The Principles of Quantum Mechanics*, 4th ed. (Oxford: Clarendon Press, 1981). In fact it is possible to derive this result from basic principles of quantum mechanics and the properties of three-dimensional space. To do so we need a prequel that we won't really introduce until the next chapter: In quantum mechanics the description of a particle or system of particles takes the form of a *wave function*—a function that depends on the position coordinates of all the particles. This wave function cannot be observed directly, which follows immediately from the fact that the wave function is *complex;* that is, it possesses both a real and an imaginary part. As the name "imaginary" suggests, such a quantity cannot be observed with any real measuring apparatus. Nevertheless, the absolute square of the wave function, equal to the sum of the squares of its real and its imaginary parts, can be interpreted as the probability density for finding the particles at particular points.

Now consider the wave function for two indistinguishable particles: $\psi(x,y)$, where x and y are the coordinates in three-dimensional space for the first and the second particle. If we interchange x and y, the absolute square of ψ cannot change, because this is the same situation, and therefore must have the same probability density. Thus, all that can happen is that the real and imaginary parts of the complex wave function are rotated into each other by an angle a, as this keeps the absolute square of ψ unchanged. This angle as a rotation between real and imaginary directions is called a *phase angle*, or for short just *phase*. We can achieve the exchange of x and y by making a rotation in space through an angle of 180 degrees, or π radians, about an axis perpendicular to and bisecting the straight line connecting x and y. If we did a rotation backward by the same amount, then we should be back where we started, so that rotation must give to the wave function a phase $-a$.

So far we have no constraint on a, but now imagine that it was *you*, looking down on a plane containing the two particles, standing between the particles and rotating them by 180 degrees counterclockwise. Someone else comes along and rotates *you*, so that you are upside down with respect to your original orientation. The locations of the two particles in your eyes are the same as ever, so this action also must give a phase rotation of the wave function; call it *b*. If you now rotate the two particles again counterclockwise about yourself from your new viewpoint, that is like rotating clockwise from your original viewpoint, and so gives a phase –*a*. Let your friend now rotate you back to right side up. This gives a phase –*b*. We have two phase rotations by +*b* and –*b*, which cancel each other out. Once we've done that, all that is left is the two counterclockwise rotations by 180 degrees you did on the two particles, giving phase 2*a*. However, if we remember that the second time you rotated the particles you were upside down, and therefore rotating the particles clockwise with respect to your original orientation, you find that the net phase was *a* – *a* = 0 (or equally well, 360 degrees, because a phase rotation by 360 degrees is equivalent to no rotation at all).

Putting all this together, we conclude that rotating twice in the same direction about the bisecting axis must give back the original ψ, which means that exchanging *x* and *y* must multiply ψ by a square root of 1, that is, either +1 or –1. The minus case gives the result that for *x* = *y*, ψ must vanish, which is the Pauli principle, while the plus case gives a wave function that is unchanged by the exchange of the coordinates of two indistinguishable particles, and that's the Bose-Einstein case. So Dirac's statement that these two possibilities are the only ones found in nature is a natural consequence of the rotational symmetry of three-dimensional space.

If we lived in a world with only two space dimensions, then this would not be so, and the angle *a* could be anything we liked (J. M. Leinaas and J. Myrheim, "On the Theory of Identical Particles," *Nuovo Cimento* B 37 [1977], pp. 1–23). Particles with this possibility available to them have been called "anyons" (Frank Wilczek, "Quantum Mechanics of Fractional-Spin Particles," *Physical Review Letters* 49 [1982], pp. 957–59). For systems built in the laboratory where motion is easy in only two rather than three dimensions, examples have been reported with evidence for particles obeying such fractional statistics.

Chapter Six: Sharks and Tigers

1. Voss-Andreae should know. He studied physics as an undergraduate in Berlin and Edinburgh, and did graduate research in Vienna with physicist Anton Zeilinger on the double-slit experiments involving the quantum interference of carbon-60 buckyball molecules. According to Voss-Andreae, while the sculpture is not directly related to the mathe-

matics of complementarity, it is related to the mathematics of quantum physics to the extent that, if you assign a chunk of energy—matter for example, or even a person—a frequency according to Planck's equation $E = hv$ and then move it, or transform it into a moving frame of reference according to the rules of special relativity, you get plane wave fronts perpendicular to the direction of motion. This is what gave Voss-Andreae the idea of the slabs (personal communication).

2. Daniel Albright, *Quantum Poetics: Yeats, Pound, Eliot, and the Science of Modernism* (Cambridge University Press, 1997), pp. 24–25.

3. John Polkinghorne, *Quantum Physics and Theology: An Unexpected Kinship* (New Haven: Yale University Press, 2007), p. 91.

4. Arthur G. Webster, *Weekly Review* 4 (1921), pp. 537–38.

5. G. E. Sutcliffe, "The Eastern School," letter to the editor, *The Times of India*, October 12, 1921, p. 11.

6. "The Quanta Theory," n.a., *Los Angeles Times*, October 20, 1922, p. II4.

7. F.C.S. Schiller, "Psychology and Logic," in *Psychology and the Sciences*, ed. William Brown (London: A & C Black, 1924).

8. In 1927, Harvard professor William Munro devoted his presidential address, at the American Political Science Association in Washington, DC, to "Physics and Politics—An Old Analogy Revisited." Bagehot's notion that democracy should be based on Newtonian mechanics was overthrown, Munro said. Political scientists had to abandon their "eighteenth-century deification of the abstract individual man," based as it is on "the atomic theory of politics—upon the postulate that all able-bodied citizens are of equal weight, volume, and value; endowed with various absolute and unalienable rights; vested with equally absolute duties; and clothed with the attribute of an indivisible sovereignty," and on the idea that government involves "a series of ultimate and fixed uniformities." Still, Munro had little to offer in the way of guidance, beyond the observation that applying quantum theory to political science might suggest "the discarding of our atomic theory of ultimate, equal, and sovereign citizens in a free state," and—embracing Bridgman's idea—counseling that if political scientists proceed "by paralleling [the natural scientist's] objectivity of attitude, and his process of operational study, the political scientist may reach that goal some day."

9. Myra Nye, "British Savant Club Speaker," *Los Angeles Times*, October 19, 1922, p. II8.

10. W. Kaempffert, "Details Concepts of Quantum Theory; Heisenberg of Germany Gives Exposition Before British Scientists," *The New York Times*, September 2, 1927, p. 6.

11. Jammer, *Conceptual Development*, p. 196.

12. Sir William Bragg, *Electrons & Ether Waves: Being the 23rd Robert Boyle Lecture, on May, 1921* (New York: Oxford, 1921).

13. The Compton effect involves the relation between angle of scattering and the change in wavelength of an X-ray photon striking an electron in an atom. The energy of the X ray is so great that the electron might as well be a free electron. Compton, using the correlation between momentum and wavelength, adding momentum conservation and energy conservation, found a formula for the shift in wavelength, and it earned him the 1927 Nobel Prize. But Compton did not know how to predict the angle of any specific photon, only the relation of the angle and the shift (*Physical Review* 21 [1923], pp. 483–502).

14. Quoted in Jammer, *Conceptual Development*, p. 171.

15. Bohr, another wave partisan, actually tried this. In 1924, he enlisted Hendrik Kramers and John Slater in a bold attempt to eradicate Einstein's idea and develop a more conventional approach that used wave theory to account for how light is emitted and absorbed, and for the photoelectric and Compton effects (N. Bohr, H. Kramers, and J. Slater, "The Quantum Theory of Radiation," *Philosophical Magazine* 47 [1924], p. 785). Like Darwin, they found that to murder Einstein's idea they had to abandon the conservation of energy, conserving it only on average. They also had to abandon any hope of a visualizable picture of the mechanics of how light is emitted and absorbed. The sacrifices of the Bohr-Kramers-Slater (BKS) theory were regarded as too extreme, not only by most physicists but even by one of its authors, Slater, who later claimed to have been coerced into signing his name. Few were surprised or even sad when, less than a year after publication, the Bohr-Kramers-Slater paper was experimentally disproven.

16. Cited in Roger Stuewer, *The Compton Effect: Turning Point in Physics* (New York: Science History Publications, 1975), p. 331.

17. J. J. Thomson, *The Structure of Light: The Fison Memorial Lecture, 1925* (Cambridge, England: Cambridge University Press, 1925), p. 15.

18. Stuewer, *The Compton Effect*, p. 331.

19. Albert Einstein, *Berliner Tageblatt*, April 20, 1924.

20. James Jeans, *Atomicity and Quanta* (Cambridge, England: Cambridge University Press, 1926), p. 62.

21. Robert P. Crease, *The Great Equations: Breakthroughs in Science from Pythagoras to Heisenberg* (New York: Norton, 2011), p. 236.

22. Werner Heisenberg, *Physics and Beyond: Encounters and Conversations* (New York: Harper and Row, 1971), p. 60.

23. Max Born, *Physics in My Generation* (New York: Pergamon Press, 1969), p. 100.

24. Pais, *Inward Bound*, p. 255.

25. See for instance Stephen Brush, "Irreversibility and Indeterminism: Fourier to Heisenberg," *Journal of the History of Ideas* 37 (1976), pp. 603–30.

26. Werner Heisenberg, *Naturwissenschaften* 14 (1926), pp. 899–904.

27. The complete correspondence, from which the following quotes are translated and drawn, is contained in A. Hermann, K. von Meyenn, and V. Weiskopf, *Wissenschaftlicher Briefwechsel mit Bohr, Einstein, Heisenberg u.a.* (New York: Springer, 1979), vol. 1.

28. See for instance Jammer, *Conceptual Development*, p. 166.

29. *Manchester Guardian*, February 14, 1935.

30. David Foster Wallace, *Everything and More: A Compact History of Infinity* (New York: Norton, 2003), p. 22.

31. John Updike, *The New Yorker*, December 30, 1985.

32. John Polkinghorne, *Quantum Physics and Theology: An Unexpected Kinship* (New Haven: Yale University Press, 2007), p. ix.

33. Teju Cole, *Open City* (New York: Random House, 2011), p. 155.

Chapter Seven: Uncertainty

1. Ray Monk, *Robert Oppenheimer: A Life Inside the Center* (New York: Doubleday, 2012), p. 142.

2. See for instance Dorothy Wrinch, "The Relations of Science and Philosophy," *Journal of Philosophical Studies* 2 (1927), pp. 153–66.

3. Craig Callendar, *The New York Times*, July 21, 2013, Online Opinion, http://opinionator.blogs.nytimes.com/2013/07/21/nothing-to-see-here-demoting-the-uncertainty-principle (last accessed July 24, 2013).

4. John Updike, *Roger's Version* (New York: Knopf, 1986), p. 168.

5. Arthur Eddington, *The Nature of the Physical World* (New York: Macmillan, 1928), p. v.

6. Eddington, *Physical World*; this and the following quotes are from the sections on the "Principle of Indeterminacy" and "A New Epistemology," pp. 220–29.

7. Eddington, *Physical World*, p. 350.

8. Eddington, *Physical World*, pp. 309, 339.

9. Gerald Holton, "Candor and Integrity in Science," in *Scientific Values and Civic Virtues*, ed. Noretta Koertge (New York: Oxford University Press, 2005), p. 87.

10. Maila L. Walter, *Science and Cultural Crisis: An Intellectual Biography of*

Percy Williams Bridgman (1882–1961) (Redwood City, CA: Stanford University Press, 1991), p. 4.

11. Bridgman to Korzybski, March 18, 1928. In Percy Williams Bridgman Papers, Harvard University, Cambridge, Massachusetts. Bridgman's article: "The New Vision of Science," *Harper's* 158, March 1929, pp. 443–54.

12. Waldemar Kaempffert, *The New York Times*, January 11, 1931.

13. "By-Products," *The New York Times*, November 3, 1929, p. E4.

14. *The New York Times*, September 5, 1936, p. 9.

15. Review of A. Eddington, "The Nature of the Physical World," *Union Seminary Review* 41 (1929), pp. 77–78.

16. Christian Wiman, *My Bright Abyss* (New York: Farrar, Straus and Giroux, 2013), p. 17.

17. This is the sense, for instance, in which President Obama referred to the uncertainty principle, as mentioned in the Introduction (*Vanity Fair*, October 2012, p. 210). "Obama then proceeded to call on every single person for his views, including the most junior people. 'What was a little unusual,' Obama admits, 'is that I went to people who were not at the table. Because I am trying to get an argument that is not being made.' The argument he had wanted to hear was the case for a more nuanced intervention—and a detailing of the more subtle costs to American interests of allowing the mass slaughter of Libyan civilians. His desire to hear the case raises the obvious question: Why didn't he just make it himself? 'It's the Heisenberg principle,' he says. 'Me asking the question changes the answer. And it also protects my decision-making.'"

18. Jon Tuttle, "How You Get That Story: Heisenberg's Uncertainty Principle and the Literature of the Vietnam War," *Journal of Popular Culture* 38 (2005), pp. 1088–98.

19. George Steiner, *Real Presences* (Chicago: University of Chicago Press, 1989). By contrast, Phillip Herring, in *Joyce's Uncertainty Principle*, sees the meaning of the principle not as defining literary interpretation but as limiting it.

20. J.W.N. Sullivan, "Science and Philosophy: Sir Arthur Eddington's New Book," *The Observer*, March 3, 1935, p. 4.

21. Others in the humanities likewise found the uncertainty principle liberating. It meant one could study and discuss things like art, values, and spiritual affairs without shame or defensiveness. "This is something of first-rate importance," wrote *Observer* columnist A. Wolf in 1929, in a series of columns on "The New Outlook in Physical Science/Universal Mechanism Abandoned/End of an Old Nightmare." "[I]t should go a

long way to set the world free for the pursuit of ideals, when economists, academic and realistic, see their idol shattered, and grasp the incongruity of making human sciences deterministic in imitation of a discredited mechanics" (*The Observer*, February 3, 1929, p. 17).

22. Wolfgang Paalen, "Art and Science," *DYN* 3 (Fall 1942), p. 8.
23. Martin Heidegger, "What Are Poets For?" in *Poetry, Language, Thought*, tr. Albert Hofstadter (New York: Harper & Row, 1971), p. 112.
24. *The New York Times*, May 25, 1930, p. 25.
25. *The New York Times*, March 27, 1931, p. 27.
26. "Religion," *The Methodist Review* 46, no. 5 (September 1930), p. 740.
27. *The Christian Science Monitor*, March 30, 1931, p. 5.
28. *The Methodist Review* 45, no. 4 (July 1929), p. 580; the *Methodist Review* is drawing much of its remarks from Eddington's book.
29. Ernest Rice McKinney, *The Pittsburgh Courier*, December 12, 1931, p. 12.
30. Quoted in Lindley, *Uncertainty*, p. 185.
31. Paalen, "Art and Science," p. 8.
32. William Barrett, *Irrational Man: A Study in Existential Philosophy* (New York: Doubleday Anchor, 1958), p. 38.
33. Barrett, *Irrational Man*, p. 40.
34. Slavoj Žižek, *The Indivisible Remainder: An Essay on Schelling and Related Matters* (New York: Verso, 1996), p. 226.
35. W. Heisenberg to W. Pauli, February 23, 1927, in A. Hermann et al., eds., *Wissenschaftlicher Briefwechsel*, pp. 377–78.
36. See Brush, "Irreversibility and Indeterminism."
37. Quoted in Brush, "Irreversibility and Indeterminism," p. 626.
38. Pais, *Inward Bound*, p. 262.
39. Interview, W. Heisenberg, February 25, 1963, in Archives for the History of Quantum Physics, p. 17, The American Institute of Physics, College Park, Maryland.

Chapter Eight: Reality Fractured

1. Shirley MacLaine, *Dancing in the Light* (New York: Bantam, 1985), p. 403.
2. James Jeans, *The Mysterious Universe* (New York: Cambridge University Press, 1930), p. 137.
3. Pais, *Inward Bound*, p. 178.
4. Arthur I. Miller, *Einstein, Picasso: Space, Time, and the Beauty that Causes Havoc* (New York: Basic Books, 2002).

5. Arthur I. Miller, "One Culture," *New Scientist*, October 29, 2005, p. 44. For more on the background to the idea of complementarity, see Gerald Holton, "The Roots of Complementarity," in *Thematic Origins of Scientific Thought: Kepler to Einstein* (Cambridge: Harvard University Press, 1988), pp. 99–145.

6. Niels Bohr, *Collected Works*, vol. 6, ed. Jørgen Kalckar (New York: North-Holland, 1985), p. 26.

7. Ibid., pp. 113–36, with the revision on pp. 580–90. This volume also has Bohr's preparatory notes and accompanying documents.

8. Ibid., p. 52.

9. Ibid.

10. Jammer, *Conceptual Development*, p. 354.

11. C. Møller, *Fysisk Tidsskrift* 60 (1962), p. 54.

12. Karl Popper, *The Logic of Scientific Discovery* (London: Routledge, 1959), p. 456.

13. J. Robert Oppenheimer, *Science and the Common Understanding* (New York: Simon & Schuster, 1954), p. 9.

14. J. Robert Oppenheimer, "Electron Theory," *Physics Today* 10 (1957), p. 20.

15. Niels Bohr, *Collected Works*, vol. 10, ed. David Favrholdt (New York: North-Holland, 1999), Editor's preface, p. v.

16. Pais, *Niels Bohr's Times*, p. 433.

17. See Bohr, *Collected Works*, vol. 10, and Pais, *Niels Bohr's Times*, pp. 438–47.

18. M. Beller, *Physics Today* 5 (September 1998), pp. 29–34.

19. Pais discusses this talk in *Niels Bohr's Times*, p. 415.

20. William L. Laurence, "Jekyll-Hyde Mind Attributed to Man," *The New York Times*, June 23, 1933, pp. 1, 13.

21. *The New York Times*, June 24, 1933, p. 12.

22. References are numerous; see for instance: J. Haas, "Complementarity and Christian Thought: An Assessment," *Journal of the American Scientific Affiliation* (1983), pp. 145–51, 203–9; R. Nadeau, *Readings from the New Book on Nature: Physics and Metaphysics in the Modern Novel* (Amherst: University of Massachusetts Press, 1981); S. Ryan, "Faulkner and Quantum Mechanics," *Western Humanities Review* 33 (1979), pp. 329–39; J. Honner, "Niels Bohr and the Mysticism of Nature," *Zygon* 17 (1982), pp. 243–53; J. Honner, "The Transcendental Philosophy of Niels Bohr, *Studies in History and Philosophy of Science* 13 (1982), pp. 1–29; R. Schlegel, "Quantum Physics and the Divine Postulate," *Zygon* 14 (1979), pp. 163–65; A. Hye, "Bertolt Brecht and Atomic Physics," *Science/Technology and the Humanities* 1 (1978), pp. 157–70; F. Falk, "Physics and the The-

atre: Richard Foreman's Particle Theory," *Educational Theatre Journal* 29 (1977), pp. 395–404.

23. Leonard Shlain, *Art and Physics: Parallel Visions in Space, Time, and Light* (New York: Morrow, 2007), p. 24.

24. See Jammer, *Conceptual Development*, section 7.2: "Complementarity."

25. Example: "Art and physics, like wave and particle . . . are simply two different but complementary facets of a single description of the world" (Shlain, *Art and Physics*, p. 24).

26. Lawrence LeShan, *The Medium, the Mystic and the Physicist: Toward a General Theory of the Paranormal* (London: Thorsons, 1974).

27. Slavoj Žižek, *Less Than Nothing: Hegel and the Shadow of Dialectical Materialism* (New York: Verso, 2012), p. 931.

28. An excellent analysis of Bohr's remark is found in Don Howard, "Who Invented the 'Copenhagen Interpretation'? A Study in Mythology," *Philosophy of Science* 71 (2004), pp. 669–82.

29. Niels Bohr, "Natural Philosophy and Human Cultures," in *Collected Works*, vol. 10, p. 89.

30. R. Crease, "The Most Beautiful Experiment," *Physics World*, May 2002, p. 17; "The Most Beautiful Experiment," *Physics World*, September 2002, pp. 17–18; "The Only Mystery: The Quantum Interference of Single Electrons," in R. Crease, *The Prism and the Pendulum: The Ten Most Beautiful Experiments in Science* (New York: Random House, 2003), pp. 191–205 (the respondent's quote is on p. 191). "Dr. Quantum" has a superb YouTube video of the double-slit experiment.

Chapter Nine: No Dice!

1. Edmund Taylor Whittaker, *From Euclid to Eddington: A Study of Conceptions of the External World* (Cambridge: Cambridge University Press, 1949), pp. 59–60.

2. See Don Howard, "'Nicht Sein Kann Was Nicht Sein Darf,' or The Prehistory of EPR, 1909–1935: Einstein's Early Worries about the Quantum Mechanics of Composite Systems," in *Sixty-Two Years of Uncertainty: Historical, Philosophical, and Physical Inquiries into the Foundations of Quantum Mechanics*, ed. Arthur Miller (New York: Springer, 1990), pp. 61–111.

3. Einstein to Hedwig Born, April 29, 1924, in *The Born-Einstein Letters*, ed. Max Born (New York: Walker & Co., 1971), p. 82.

4. Quoted in Lindley, *Uncertainty*, p. 133.

5. *Nature*, March 26, 1927, p. 467.

6. Guido Bacciagaluppi and Antony Valentini, *Quantum Theory at the Cross-*

roads: Reconsidering the 1927 Solvay Conference (Cambridge: Cambridge University Press, 2009), pp. 234, 237.

7. Ibid., p. i.
8. Ibid., Ch. 2.
9. Ibid., pp. 440ff.
10. Ibid., 175–76.
11. Quoted in Bacciagaluppi and Valentini, *Quantum Theory*, p. 442.
12. Bacciagaluppi and Valentini, *Quantum Theory*, pp. 247–82.
13. Leon Rosenfeld, quoted in Pais, *Niels Bohr's Times*, pp. 446–47.
14. A. Einstein, R. Tolman, and Boris Podolsky, "Knowledge of the Past and Future in Quantum Mechanics," *Physical Review* 37 (1931), pp. 780–81.
15. "Einstein Affirms Belief in Causality," *The New York Times*, March 17, 1931, p. 13.
16. "Conservative Einstein," *The New York Times*, March 18, 1931.
17. William L. Laurence, "Einstein Offers New View of Mass-Energy Theorem," *The New York Times*, December 29, 1934, pp. 1, 7.
18. A. Einstein, B. Podolsky, and N. Rosen, "Can Quantum-Mechanical Description of Physical Reality Be Considered Complete?" *Physical Review* 47 (1935), pp. 777–80.
19. Paul Schilpp, ed., *Albert Einstein, Philosopher-Scientist: The Library of Living Philosophers*, vol. VII (Chicago: Open Court, 1949), p. 666.
20. Russell McCormmach, *Night Thoughts of a Classical Physicist* (New York: Avon, 1982).
21. Robert P. Crease and Charles C. Mann, "Interview: John Bell," *Omni* (May 1988), pp. 85–92, 121.
22. Jeffrey Bub, "Von Neumann's 'No Hidden Variables' Proof: A Re-Appraisal," arXiv:1006.0499.
23. David Bohm, *Physical Review* 85 (1952), pp. 166, 189.
24. John Bell, *Speakable and Unspeakable in Quantum Mechanics* (Cambridge: Cambridge University Press, 1987), pp. 1–13.
25. Ibid., pp. 14–21.
26. J. Clauser, M. Horne, A. Shimony, and R. Holt, "Proposed Experiment to Test Local Hidden-Variable Theories," *Physical Review Letters* 23 (1969), pp. 880–84.
27. Our favorite concise explanation of Bell's theorem, which we distribute to the class, is N. David Mermin, "Is the Moon There When Nobody Looks? Reality and the Quantum Theory," in *Physics Today* 38 (April 1985), pp. 38–47, with a different version appearing as "Quantum Mysteries for Anyone," in *Philosophy of Science and the Occult*, 2nd ed., ed. P. Grim (Albany: State University of New York Press, 1990), pp. 315–25.
    While we don't know how to give a brief, accessible account of Bell's

theorem in its original form, we *can* give a good way to picture how entanglement works. First, let us note that *all* the weird properties of quantum mechanics, superposition, uncertainty, etc., including entanglement, really are wave effects that can be seen as completely normal for classical waves, and only become strange when we use quantum mechanics to apply waves to predictions about particles.

Consider an experiment in which a photon beam is sent through a "parametric down-conversion" device that takes each photon and converts it to two photons with half the frequency and therefore half the energy. These two photons automatically are correlated in momentum. So not surprisingly if one is observed with a certain momentum to the left, the other will have the same momentum to the right. That doesn't seem terribly surprising, but now let's imagine a different experiment, in which the first photon passes through a mask with openings that make up, for example, the letters U-M-B-C (T. B. Pittman, Y. H. Shih, D. V. Strekalov, and A. V. Sergienko, "Optical Imaging by Means of Two-Photon Quantum Entanglement," *Physical Review* A 52 [1995], p. R3429). The photon on the other side is collected without any shapes being imposed on its allowed paths. Once enough photons have passed to the left to make a visible pattern U-M-B-C on a collection screen, one may look at the pattern of the other photons that had no constraint on their paths. Lo and behold, these photons, each coincident in time with one of the left-moving photons, generate a pattern that again is U-M-B-C! This is precisely the thing that EPR protested: The two particles are linked in momentum, but also linked in position, so that they clearly are entangled, because for either particle separately we could not simultaneously measure position and momentum. The link in position is a subtle one. It's not that the coincident photons end up in exactly corresponding positions. Rather, the constraint on the left-moving photons creates a 'hologram' for the right-moving photons, so that the cumulative effect for all the right-moving photons (with, as mentioned, each coincident in time with one of the detected left-moving photons) reproduces the image forced on the left-moving ones.

The proof that this entanglement is a wave phenomenon comes from an analogue experiment that uses an intense beam of laser light that amounts to a classical wave. Such a wave can be split using a half-silvered mirror that reflects half the intensity and transmits the other half. This time, the momenta parallel to the mirror surface (for the classical wave, the analogous quantities are called "wave numbers") must be equal, creating an obvious correlation in momentum, but instead one may look at a shape on the one side and see it reproduced on the other. EPR would have had no objection to this, because for waves such a relationship is an "expected" form of reality (Ryan S. Bennink, Sean J. Bentley, and Robert W. Boyd, "'Two-Photon' Coincidence Imaging with a Classical Source," *Physical Review Letters* 89 [2002], p. 113601).

Chapter Ten: Schrödinger's Cat

1. See for instance Howard, "Who Invented the 'Copenhagen Interpretation'?"
2. E. Schrödinger to A. Einstein, June 7, 1935. This letter is quoted in Walter Moore, *Schrödinger: Life and Thought* (New York: Cambridge University Press, 1989), p. 304.
3. Ibid.
4. Ibid., p. 305.
5. E. Schrödinger, "Die gegenwärtige Situation in der Quantenmechanik," *Naturwissenschaften* 23 (1935), pp. 807–12; 823–32, 844–49. For the English translation see John D. Trimmer, "The Present Situation in Quantum Mechanics," *Proceedings of the American Philosophical Society* 124 (1980), pp. 323–38, reprinted in J. A. Wheeler and W. H. Zurek, eds., *Quantum Theory and Measurement* (Princeton: Princeton University Press, 1983), p. 152; also online: http://www.tu-harburg.de/rzt/rzt/it/QM/cat.html#sect5.
6. For a good discussion see Howard, "Nicht Sein Kann."
7. Stephen Brush, "The Chimerical Cat: Philosophy of Quantum Mechanics in Historical Perspective," *Social Studies of Science* 10 (1980), pp. 393–447.
8. For Schrödinger's references see Karl Meyenn, *Eine Entdeckung von ganz außerordentlicher Tragweite: Schrödingers Briefwechsel zur Wellenmechanik und zum Katzenparadoxon* (Berlin: Springer, 2011).
9. As in Anthony J. Leggett, "Quantum Mechanics: Is It the Whole Truth?" in *Visions of Discovery: New Light on Physics, Cosmology, and Consciousness*, ed. Raymond Y. Chiao et al. (Cambridge: Cambridge University Press, 2011), pp. 171–183; see also A. J. Leggett, *Journal of Physics* A 40 (2007), p. 3141.
10. Gary Zukav, *The Dancing Wu Li Masters* (New York: William Morrow, 1979), p. 86.
11. As a topic of popular culture, Schrödinger's cat reached maturity in the 1980s. In 1984, for instance, the scientist and science writer John Gribbin published *In Search of Schrödinger's Cat: Quantum Physics and*

Reality, whose popularity inspired a sequel: *Schrödinger's Kittens and the Search for Reality* (1996). Other nonfiction works discussing Schrödinger's image include *Who's Afraid of Schrödinger's Cat? An A-to-Z Guide to All the New Science Ideas You Need to Keep Up with the New Thinking* (Ian Marshall and Danah Zohar, 1997), and *Schrödinger's Machines: The Quantum Technology Reshaping Everyday Life* (Gerald J. Milburn, 1997). Novels of the past few decades incorporating the theme include *Schrödinger's Cat Trilogy* (Robert Anton Wilson, 1979); *Schrödinger's Baby: A Novel* (H. R. McGregor, 1999); *Schrödinger's Ball* (Adam Felber, 2006); and *Blueprints of the Afterlife* (Ryan Boudinot, 2012). Short stories include Ursula K. Le Guin's "Schrödinger's Cat" (1974) and F. Gwynplaine MacIntyre's "Schrödinger's Cat-Sitter" (2001). "Schrödinger's Cat" is the name of a musical group, *Schrödinger's Mouse* the name of a science-fiction magazine. TV shows that have referred to the cat include "Stargate," "Futurama," "CSI," and an audio episode of "Doctor Who." John Lennon's son Sean, as well as a Ukrainian rock group, have released songs with "Schrödinger's cat" in the title. The cat appears on YouTube and in numerous video games. There is even a Wikipedia page devoted to "Schrödinger's cat in popular culture."

12. Thanks to the (then) postdoctoral students from Bristol and Oxford who sent these to us: David Leigh, Gavin Morley, Denzil Rodrigues, and Jamie Walker.

13. Lily Tuck, *I Married You for Happiness* (New York: Atlantic Monthly Press), pp. 126–27.

14. http://www.youtube.com/watch?v=itQVDA6_TME.

15. Steve Field, "Quantum Socks," *New Scientist*, June 23, 2012, p. 31.

16. J. R. Friedman, V. Patel, W. Chen, S. K. Tolpygo, and J. E. Lukens, "Quantum Superposition of Distinct Macroscopic States," *Nature* 406 (2000), pp. 43–46.

17. "How Fat is Schrödinger's Cat," *Physics World*, April 25, 2013, http://physicsworld.com/cws/article/news/2013/apr/25/how-fat-is -schrodingers-cat, accessed October 1, 2013; also "Comments," *Physics World*, June 2013, p. 21.

Chapter Eleven: Rabbit Hole

1. Movies depicting alternate versions of reality in ways that serve yet other purposes include *It's a Wonderful Life* (1946), *Groundhog Day* (1993), *Sliding Doors* (1998), and *Back to the Future* (1985). See also alternate histories, as in *Steamboy* (2004), a Japanese film about an alternate Europe, and *Difference Engine*, a novel by William Gibson and Bruce Sterling (1992).

Books with alternate worlds include Ray Bradbury's *Martian Chronicles* (1950, the third expedition) and Diana Jones's *The Lives of Christopher Chant* (1988). Movies include *The Girl Who Leapt Through Time*, *The Source Code*, and *Lake House*.

2. For a biography of Everett, see Peter Byrne, *The Many Worlds of Hugh Everett III: Multiple Universes, Mutual Assured Destruction, and the Meltdown of a Nuclear Family* (New York: Oxford University Press, 2010), and the review by R. Crease in *Nature* 465 (2010), pp. 1010–11. For a collection of papers, see Bryce S. DeWitt and Neill Graham, eds., *The Many-Worlds Interpretation of Quantum Mechanics* (Kathmandu: Five Mile Mountain Press, 1980).

3. Byrne, *Many Worlds*, p. 11.

4. Hugh Everett, III, "'Relative State' Formulation of Quantum Mechanics," *Reviews of Modern Physics* 29 (1957), pp. 141–49, reprinted in DeWitt and Graham, *Many-Worlds Interpretation*.

5. DeWitt and Graham, *Many-Worlds Interpretation*, p. v.

6. Ibid., p. xii.

7. DeWitt and Graham, *Many-Worlds Interpretation*, p. 161.

8. Kurt Vonnegut, *Slaughterhouse-Five* (New York: Delacorte Press, 1969).

9. *Doctor Who*, Episode 5, Series 2, 10th Doctor.

10. "On View: It's Fashionable to Take a Trip to Another Universe," *The New York Times*, July 26, 2011.

11. Michael Douglas, *Journal of High Energy Physics* 0305 (2003), p. 46.

Chapter Twelve: Saving Physics

1. Howard, "Who Invented the 'Copenhagen Interpretation'?"

2. Quoted in Howard, "Who Invented the 'Copenhagen Interpretation'?" p. 676.

3. Jack Sarfatti and Fred Alan Wolf, personal communication, March 17, 2013.

4. David Kaiser, *How the Hippies Saved Physics: Science, Counterculture, and the Quantum Revival* (New York: Norton, 2011), p. 119.

5. Fritjof Capra, *The Tao of Physics: An Exploration of the Parallels Between Modern Physics and Eastern Mysticism* (Berkeley: Shambhala Publications, 1976), p. 298.

6. D. Harrison, "What You See Is What You Get!" *American Journal of Physics* 47 (1979), pp. 576–82; "Teaching the Tao of Physics," *American Journal of Physics* 47 (1979), pp. 779–83.

7. Zukav, *The Dancing Wu Li Masters*.

8. Buffalo: State University of New York Press, 1984.

9. The textbook that briefly mentioned Bell's theorem was Kurt Gottfried, *Quantum Mechanics: Fundamentals* (W. A. Benjamin, 1966). The first quantum mechanics textbook that Kaiser has found that devotes any attention to Bell's theorem was Sakurai's 1985 textbook *Modern Quantum Mechanics*; i.e., Bell's theorem did not enter mainstream physics textbooks until after the Fundamental Fysiks Group had left its impact.

10. See for instance Martin Gardner, "Quantum Theory and Quack Theory," in the *New York Review of Books* 26 (May 17, 1979); "Magic and Paraphysics," *Technology Review* 78 (June 1976), pp. 42–66.

11. Paulo Coehlo, *Brida: A Novel* (New York: Harper Collins 2008), pp. 73–74.

12. Thanks to George Clinton for this remark.

13. Abraham Pais, *Subtle Is the Lord: The Science and the Life of Albert Einstein* (New York: Oxford, 2005), p. 1.

14. Marshall Spector, "Mind, Matter, and Quantum Mechanics," in *Philosophy of Science and the Occult*, 2nd ed., ed. P. Grim (Albany: State University of New York Press, 1990), p. 344.

15. Consider the following remarks from Brush, *Chimerical Cat*, p. 422: Born: "I personally like to regard a probability wave, even in 3N-dimensional space, as a real thing, certainly as more than a tool for mathematical calculations." Weisskopf: "The fact that atomic phenomena are hard to catch and hard to describe does not make them less real."

Conclusion: The Now Moment

1. See for instance Lillian Hoddeson, "The Entry of the Quantum Theory of Solids into the Bell Telephone Laboratories, 1925–40: A Case-Study of the Industrial Application of Fundamental Science," *Minerva*, 1980, pp. 422–47.

2. For our class, we often put many of these terms and images into a table:

TERM	ORIGIN	METAPHORICAL APPLICATION
Quantum (1900)	Explanation of black-body radiation	Discreteness
Quantum leap (1913–14)	Discontinuous state transition	Huge transition of anything
Randomness (1916–17)	Atomic light emission and absorption	Order from disorder, chaos

Wave-particle duality (1920s)	Behavior of quantum phenomena	Schizophrenia
Uncertainty principle (1927)	$\Delta p \Delta q \geq \hbar/2$	Unpindownability, observer effect, randomness, free will, consciousness affects reality, unpicturability . . .
Complementarity (1927)	Two features of a phenomenon (e.g., waves and particles) can be necessary yet mutually exclusive	The fact that a phenomenon may exhibit paradoxical or incompatible behaviors, or may appear different depending on the approach taken by the observer
Schrödinger's cat (1935)	A joke mocking the fact that the wave function implies something may have indeterminate existence until "observed"; a realist critique of the Copenhagen Interpretation	An image or joke, sometimes sadistic, about indeterminate existence, or life and death, or the odd behavior of cats
Parallel worlds (1957)	An interpretation of quantum mechanics, opposed to the Copenhagen Interpretation, in which wave functions do not "collapse." No interaction between worlds is possible.	A fictional conceit—or reality promised by quacks and shamans—in which parallel worlds exist and travel between them is possible. "I should have done that other thing"-ing.
Multiverses (2004)	An application of string theory in which quantum fluctuations create a cornucopia of "baby universes" which grow large through inflation but do not overlap with one another	Our universe as only one of an unlimited number of universes

3. Dobbs and Jacob, *Newton*, p. 123.

Acknowledgments

We have an astonishingly broad spectrum of acknowledgments to make. The two of us first met about ten years ago because we had independently submitted proposals for the same grant, and the Provost asked us to work together. Meanwhile, a freshman at Stony Brook named Maaneli Derakhshani was wandering the halls of both our departments in hope of finding a philosopher and a physicist who would lead a course on the philosophy of quantum mechanics. We were the only pair who took him seriously, and started thinking about what might make sense and be accessible to both philosophy and physics undergraduate students. We knew there was plenty of material on the philosophy, history, and science of quantum mechanics—the field hardly needed more—but the idea of looking at the impact of quantum mechanics on general culture (and vice versa) struck us as appealing. The course more than fulfilled these hopes. We called our course "The Quantum Moment," inspired by a 2004 exhibition at the New York Public Library, curated by the science historian Mordechai Feingold, called "The Newtonian Moment." The two of us are grateful not only to Stony Brook University for allowing us to teach such an innovative course, but also to the University administrators (Alissa Betz in the Philosophy Department, Pam Burris and Nathan Leoce-Shappin in the Physics Department) for struggling, year after year, with the mechanics of fitting a cotaught course into a demanding curriculum.

In the first iteration of the course, in spring 2006, we were lucky enough to have not only talented and eager students but also several senior faculty participating, including physicist Harold Metcalf, philosophers Patrick Grim and Lee Miller, and biochemist Paul Bingham. Each year, the class took a field trip to the Stony Brook Physics Department's Laser Teaching Center, supervised by Metcalf and the LTC's executive director John Noé. We have also had a string of eminent guest speakers, some who appeared in person, others by Skype, a few both. These included Paul Forman, David Kaiser, Nobuko Miyamoto, and Gino Segrè. These speakers added to the insights for our students, and also to our own perspectives on this rich subject.

In summer 2012, one of us (Crease) taught a miniature version of the course at Stony Brook's NY–St. Petersburg Institute of Linguistics, Cognition and Culture (NYI) in St. Petersburg, Russia, and we are grateful to its codirector, Stony Brook Professor John Frederick Bailyn, as well as to the numerous enthusiastic students who took the course and provided us with a Russian perspective. One of us (Goldhaber) gave a talk related to our course at the 2012 Bridges: Mathematical Connections in Art, Music, and Science, at Towson University, Maryland, July 25–29; the other of us (Crease) gave a talk at the 10th Gathering for Gardner (G4G10) in Atlanta in March 2012; and we gave a joint talk on the course at Stony Brook's Humanities Institute in September 2012.

. One of us (Crease) writes a monthly column, "Critical Point," for *Physics World* magazine, an exceptionally edifying and enlightening magazine published by the Institute of Physics; editor Matin Durrani ensures that it is demanding to write for and enjoyable to read. We are grateful not only to him but to the hundreds of people who responded to those columns. Some of the material in this book appeared first in those columns, including parts of Chapters 1 ("The Quantum Moment," March 2013), 3 ("Fruitloopery," February 2012), 5 ("Identity Physics," January 2013), 9 ("No-way Physics," July 2007), 10 ("The Cat That Never Dies," April 2013), and 11 ("Otherworldly Tales," December 2011); other material was drawn from columns of December 2001 ("Too Confident About Uncertainty"), May 2008 ("The Bohr Paradox"), and September 2008 ("Quantum of Culture"). I am grateful for permission to revise and adapt this material.

We are indebted to our editor at Norton, Maria Guarnaschelli, for her insightful readings of early versions of the manuscript and for her patience while we tried to reshape the manuscript more coherently. We are also indebted to Mitchell Kohles, for helping to steer the manuscript through; to managing editor Nancy Palmquist and production manager Devon Zahn; to Julie Sherrier; and to Carol Rose, the copyeditor, whose numerous suggestions considerably improved the writing.

We are also indebted to John Green and David Levithan, who allowed us to reprint the long and interesting discussion about Schrödinger's cat between Will (well, one of them) and Jane from *Will Grayson, Will Grayson*. We are grateful to the Harvard University Library, where we used the Percy Williams Bridgman collection, and to the Columbia University Library and Stony Brook University library, which we used extensively. We are also indebted for various insights to Edward S. Casey, Patrick Grim, George W. Hart, Don Ihde, the late John H. Marburger III, Eduardo Mendieta, Hal

Metcalf, Lee Miller, Robert C. Scharff, and Marshall Spector. Many thanks to Roy Glauber for allowing us to reprint his wonderful picture of Pauli and to quote from his description of how the picture came to be. Chi Ming Hung, systems manager at the C. N. Yang Institute for Theoretical Physics, drew several of the diagrams. Thanks to the ever-surprising xkcd for his imaginative cartoons and his generosity in allowing people to share them. One of us (Crease) could not have done the book without the love and support of his wife Stephanie, who read many versions of the manuscript and contributed significantly to its clarity and content—and helped edit many of the columns that preceded the manuscript. He also thanks his son Alexander, who worked on some of the diagrams, and his daughter India, for their love and support. Goldhaber thanks his wife Suzan Goldhaber for her love, encouragement, and advice, and their children David Goldhaber-Gordon and Sara Goldhaber-Fiebert and their families, who have made life so special and rewarding before and during the work on this book. In addition, David has made insightful comments on our text. Finally, we would like to thank each other for navigating six years of vastly different sensibilities, styles, and schedules in a way that made each class seem novel and fresh.

Credits

p. 2: Courtesy Juan Mesa.

p. 5: www.CartoonStock.com.

p. 7: © by Alison Bechdel. Reprinted by permission of Houghton Mifflin Harcourt Publishing Company. All rights reserved.

p. 8: xkcd.com.

p. 12: ©Tate, London 2013.

p. 20: Courtesy the Sir Isaac Newton Pub, 84 Castle Street, Cambridge, England.

p. 31: Courtesy Chi Ming Hung.

p. 35: Emilio Segrè Visual Archives, Gift of Hans Kangro.

p. 37: Courtesy Franklin W. Olin College of Engineering.

p. 38: Photographie Benjamin Couprie, Institut International de Physique Solvay, courtesy AIP Emilio Segrè Visual Archives.

p. 41: Courtesy AIP Emilio Segrè Visual Archives.

p. 50: Courtesy Chi Ming Hung.

p. 57: Niels Bohr Archive, Blegdamsvej 17, DK-2100 Copenhagen Ø, Denmark, courtesy AIP Emilio Segrè Visual Archives.

p. 62: Courtesy *Saturday Morning Breakfast Cereal* (SMBC), webcomic by Zachary Alexander Weinersmith. Zach Weinersmith, zach@smbc-comics.com.

p. 64: www.CartoonBank.com. Mark Thompson/The New Yorker Collection.

p. 70: Photograph by Stephen White, London © the artist.

p. 72: Courtesy Tony Zurlo.

p. 86: © 1962 Schroder Music Co. (ASCAP), Renewed 1990, Used by Permission. All rights reserved.

p. 90: xkcd.com.

p. 95: AIP Emilio Segre Visual Archives, Gift of Kameshwar Wali and Etienne Eisenmann.

p. 102: Courtesy Roy Glauber.

p. 110: Courtesy Julian Voss-Andreae.

p. 111: xkcd.com.

p. 119: Courtesy Sidney Harris, ScienceCartoonsPlus.com.

p. 131: Courtesy Chi Ming Hung.

p. 134: Courtesy Chi Ming Hung.

p. 137: Courtesy Sidney Harris, ScienceCartoonsPlus.com.

p. 139: Courtesy Universal Uclick.

p. 139: www.domaineboyar.com.

p. 142: Courtesy AIP Emilio Segrè Visual Archives, Gift of Subrahmanyan Chandrasekhar.

p. 144: Photo by Bachrach.

p. 152: Courtesy Universal Uclick.

p. 153: Courtesy Jeroen Vanstiphout, www.kartoen.be.

p. 166: xkcd.com.

p. 166: Courtesy Alexander Crease.

p. 170: © 2013 Artists Right Society (ARS), New York/ADAGP, Paris. Image: ©SMK Photo

p. 206: © Peter Menzel, www.menzelphoto.com.

p. 211: From *Will Grayson, Will Grayson* by John Green and David Levithan, copyright © 2010 by John Green and David Levithan. Used by permission of Dutton Children's Books, a division of Penguin Group (USA) Inc.

p. 220: Michael Ramus, Courtesy *Physics Today*.

p. 223: xkcd.com.

p. 224: icanbarelydraw.com CC BY-NC-ND 3.0.

p. 233: "Another Earth,"™ and ©Twentieth Century Fox Film Corporation.

p. 242: Courtesy Nature Publishing Group.

p. 245: Courtesy Ryan Lake.

p. 246: Lee Lorenz/The New Yorker Collection/www.cartoonbank.com.

p. 255: © Donna Coveney.

p. 273: xkcd.com

Index

Page numbers in *italics* refer to illustrations.
Page numbers beginning with 281 refer to endnotes.

About the Authors

Robert P. Crease is a professor in the Department of Philosophy at Stony Brook University. He is Co-Editor-in-Chief of *Physics in Perspective*, and writes a monthly column, "Critical Point," for *Physics World* magazine. He is a Fellow of the Institute of Physics (IOP) in London, and of the American Physical Society (APS) in the United States. He has written, co-written, translated, or edited over a dozen books. His previous books include *World in the Balance: The Historic Quest for an Absolute System of Measurement*; *The Great Equations: Breakthroughs in Science from Pythagoras to Heisenberg*; and *The Prism and the Pendulum: The Ten Most Beautiful Experiments in Science*. His articles and reviews have appeared in the *The Atlantic Monthly*, the *New York Times*, the *Wall Street Journal*, *Science*, *New Scientist*, *American Scientist*, and other scholarly and popular publications. He enjoys playing "Dr. Matrix" with his son at conferences in honor of the philosopher and recreational mathematician Martin Gardner, and can't survive the week without an African dance class. Web page: www.robertpcrease.com. He lives in New York City.

Alfred Scharff Goldhaber, professor in the Yang Institute for Theoretical Physics at Stony Brook University and fellow of the American Physical Society, represents the second of three physics generations in his family. Collaborations with his parents led to what may have been the first mother-son publications in physics, on oscillations of atomic nuclei. He collaborated with his father on a hundredth-birthday article in *Physics Today* about properties of neutrinos. He works on magnetic monopoles, elementary particles, nuclei, condensed matter, astrophysics, and cosmology, and has authored numerous review articles about particular research topics in physics. He enjoys "hindsight heuristics," asking why people made discoveries later than they might have: understanding this better could aid future discoveries. Besides "The Quantum Moment" with Robert P. Crease, he created a course, "Physical and Mathematical Foundations of Quantum Mechanics," introducing the subject to undergraduate students. His general lectures include "Ampère, Faraday, and the How of Fundamental Physics," and "Einstein the Radical versus Einstein the Conservative."